String Theory Step by Step

An Introduction to String Theory for the Non-Physicist

James Young, PhD

PREFACE

Welcome, curious reader! You've just opened the door to one of the most fascinating and mind-bending realms of modern physics: string theory. Don't worry if you don't have a strong background (or any background) in physics or string theory – this book is designed to be your friendly guide through this complex but captivating subject.

"String Theory Step by Step" is written with you in mind – the non-physicist who's intrigued by the big questions about our universe but might feel a bit lost in the sea of scientific jargon, or simply where to start in understanding such a subject. My goal is to take you on a journey, step by step, through the theoretical framework of tiny vibrating strings that might just be the fundamental building blocks of everything we see around us.

You might be wondering, "Why should I care about string theory?" Well, imagine if we could explain all the forces of nature – gravity, electromagnetism, and the nuclear forces – with one elegant theory. That's the promise of string theory. It's an ambitious attempt to unify our understanding of the universe, from the tiniest subatomic particles to the vast expanse of the cosmos.

Throughout this book, we'll explore the history of string theory, its fundamental concepts, and how it relates to other areas of physics like quantum mechanics and relativity. We'll venture into mind-bending ideas like extra dimensions and the multiverse. Don't worry if these concepts sound alien – we'll approach them gradually and with plenty of explanations along the way.

One of the most exciting aspects of string theory is how it challenges our perception of reality. We'll discuss how the mathematics behind string theory suggests that our universe might have more dimensions than the three spatial dimensions we're familiar with. We'll also go into the idea that everything in the universe, at its most fundamental level, might be composed of tiny vibrating strings of energy.

But this book isn't just about abstract theories. We'll also look at the practical side of string theory – how scientists are trying to test its predictions, the technology used in these experiments, and the challenges they face. We'll explore both the promise and the criticisms of string theory, giving you a balanced view of this controversial field.

As we progress through the chapters, you'll find that each builds upon the last. At the same time, each section and chapter can also stand on its own for those who wish to pick and choose. But don't worry – I've made every effort to keep the language clear and accessible. Where mathematical concepts are necessary, I'll introduce them gently and explain their significance.

By the end of this book, you won't necessarily be a string theory expert – that takes years of advanced study – but you will have a solid grasp of what string theory is, why it's important, and how it might shape our understanding of the universe in the years to come. You'll be able to follow discussions about string theory in the media and perhaps even engage in some fascinating conversations about the nature of reality.

Whether you're a student, a science enthusiast, or simply someone with a curious mind, I invite you to turn the page and step into the wonderful world of string theory.

TOPICAL OUTLINE

Chapter 1: What is String Theory?

- The Basics of String Theory
- Why Strings?
- History of String Theory
- Particles vs. Strings
- String Theory in a Nutshell
- Dimensions Beyond the Observable Universe
- The Role of Mathematics in String Theory
- String Theory and Quantum Mechanics
- String Theory vs. Classical Physics
- Future of String Theory

Chapter 2: The Birth of String Theory

- Early Theories and Discoveries
- Key Scientists and Contributors
- Evolution from Particle Physics
- The First String Revolution
- Why a Unification Theory in Physics is So Difficult to Achieve

Chapter 3: Fundamental Concepts of String Theory

- Strings and Branes
- Open and Closed Strings
- Vibrations and Modes
- String Interactions

Chapter 4: The Mathematical Foundations

- Basic Mathematical Concepts & Techniques
- Complex Numbers and Their Role
- Symmetries in String Theory
- Calabi-Yau Manifolds

Chapter 5: String Theory and Quantum Mechanics

- Unifying Principles
- Quantum Gravity
- Quantum Fields and Strings
- Conformal Field Theory

- The Holographic Principle

Chapter 6: Dimensions and The Multiverse

- Understanding Higher Dimensions
- String Theory's Extra Dimensions
- Compactification
- The Concept of the Multiverse

Chapter 7: String Theory and Relativity

- General Relativity Overview
- String Theory's Compatibility with Relativity
- Bridging Quantum Mechanics and Relativity
- Gravitons in String Theory

Chapter 8: Types of String Theory

- Bosonic String Theory
- Superstring Theory
- M-Theory
- Duality Relationships

Chapter 9: Supersymmetry

- Basics of Supersymmetry
- Supersymmetry in String Theory
- Superpartners
- Experimental Challenges

Chapter 10: String Theory in Modern Physics

- Impact on Cosmology
- Black Holes and Strings
- String Theory in Particle Physics
- String Theory vs. Standard Model
- AdS/CFT Correspondence

Chapter 11: Experimental Approaches

- Collider Experiments
- Astrophysical Observations
- The Search for Supersymmetric Particles
- Technological Limitations

Chapter 12: Criticisms and Controversies

- Theoretical Criticisms
- Philosophical Debates
- String Theory vs. Loop Quantum Gravity

Chapter 13: Applications of String Theory

- Technology and String Theory
- Cosmological Applications
- Potential Future Applications
- String Theory in Other Scientific Fields

Chapter 14: The Future of String Theory

- Next Steps in Research
- Competing Theories and Approaches to String Theory
- Potential Breakthroughs
- Unification with Other Theories
- Implications for Our Understanding of the Universe

Appendix

- Terms and Definitions
- String Theory Timeline

Afterword

TABLE OF CONTENTS

CHAPTER 1: WHAT IS STRING THEORY?

The Basics of String Theory

String theory proposes that the fundamental building blocks of the universe are not point-like particles but tiny, vibrating strings. These strings are incredibly small, on the order of the Planck length, which is about 10^{-35} meters. In string theory, the different particles we observe in nature are just different vibrational modes of these strings, much like different notes on a guitar string.

One of the core ideas of string theory is that it requires more than the familiar three dimensions of space and one dimension of time. To be consistent, string theory necessitates additional spatial dimensions. In the simplest versions, there are ten dimensions: nine of space and one of time. In some variants, like M-theory, there are eleven dimensions. These extra dimensions are not something we see in our everyday lives because they are compactified, meaning they are curled up in such a way that they are too small to detect directly.

String theory emerged as an attempt to unify the forces of nature. The four fundamental forces—gravity, electromagnetism, the strong nuclear force, and the weak nuclear force—have traditionally been described by different theories. Electromagnetism and the weak nuclear force are unified under the Standard Model of particle physics, but gravity has remained elusive. String theory naturally incorporates gravity, offering a potential way to reconcile quantum mechanics and general relativity.

At its heart, string theory replaces the point-like particles of particle physics with one-dimensional objects. The string can be open, like a piece of thread, or closed, forming a loop. The type of string and the way it vibrates determine the particle's properties, such as mass and charge. This is a radical departure from traditional physics, which treats particles as points without internal structure.

The mathematics of string theory is complex and requires advanced tools from geometry and topology. One key mathematical concept in string theory is the idea of a "brane." Branes are multi-dimensional objects that generalize the notion of a string. For example, a one-brane is a string, a two-brane is a membrane, and so on. These branes can have various dimensions and can interact with strings in interesting ways.

String theory also predicts the existence of supersymmetry, a symmetry that relates particles of different spins. In a supersymmetric theory, every particle has a superpartner with a spin that differs by half a unit. Supersymmetry has not been observed experimentally, but it is a crucial part of string theory because it helps cancel out certain mathematical infinities that would otherwise make the theory inconsistent.

One of the most fascinating aspects of string theory is its potential to describe all fundamental particles and forces in a single framework. However, this same feature makes it difficult to test. The strings are so small that detecting them directly with current technology is impossible. Experiments like those conducted at the Large Hadron Collider (LHC) search for indirect evidence, such as the production of supersymmetric particles.

Despite its elegance, string theory faces challenges. One of the main criticisms is the lack of unique predictions. The theory has a vast landscape of possible solutions, known as the "string landscape," each corresponding to a different vacuum state with its own physical properties. This makes it difficult to pinpoint which version of string theory, if any, describes our universe.

String theory has also led to important insights in mathematics and theoretical physics. For example, it has deep connections to various branches of mathematics, such as algebraic geometry and number theory. These connections have inspired new mathematical techniques and solved long-standing problems.

In recent years, string theory has branched out into other areas of physics. It has provided tools to study black holes and the nature of spacetime itself. One notable success is the holographic principle, which suggests that a higher-dimensional space can be described by a lower-dimensional boundary theory. This idea has profound implications for our understanding of gravity and quantum mechanics.

While string theory remains a work in progress, it has already transformed how physicists think about the fundamental nature of reality. Its ambitious goal of unifying all forces and particles into a single framework continues to inspire and challenge scientists. Even if string theory does not ultimately provide the final theory of everything, it has already enriched our understanding of the universe and opened new avenues for exploration.

Why Strings?

String theory was developed to address several significant issues in physics, particularly the unification of the fundamental forces and the reconciliation of quantum mechanics with general relativity. Here's why the concept of strings became central to this endeavor:

1. **Problem with Point Particles:** In traditional quantum field theory, particles are treated as point-like objects. However, when physicists attempted to include gravity in this framework, they encountered severe

mathematical problems. The equations produced infinities that couldn't be easily resolved. String theory solves this by replacing point particles with one-dimensional strings. When strings interact, the interactions are spread out over a small but finite length, avoiding the problematic infinities.

2. **Unification of Forces:** One of the holy grails of physics is to unify all four fundamental forces: gravity, electromagnetism, the weak nuclear force, and the strong nuclear force. The Standard Model successfully unifies electromagnetism, the weak force, and the strong force, but gravity has remained separate. String theory naturally includes gravity. In fact, one of the vibrational modes of a string corresponds to the graviton, the hypothetical quantum particle that mediates the force of gravity.

3. **Diversity of Particles:** In string theory, different particles arise from different vibrational states of the strings. Just as different frequencies of a vibrating guitar string produce different musical notes, different vibrational patterns of a string produce different particles. This elegant mechanism provides a unified way to describe the variety of particles observed in nature.

4. **Supersymmetry:** String theory predicts a symmetry known as supersymmetry, which posits that each particle has a superpartner with different spin characteristics. Supersymmetry helps cancel out certain mathematical infinities that arise in quantum field theory, making the equations more manageable. While supersymmetry has not yet been observed experimentally, it remains a key feature of string theory.

5. **Extra Dimensions:** String theory requires more than the familiar three spatial dimensions and one-time dimension. The theory is consistent in a higher-dimensional space, typically 10 or 11 dimensions. These extra dimensions are compactified, meaning they are curled up so small that they are undetectable at human scales. These additional dimensions allow for the complex interactions needed to describe all fundamental forces within a single framework.

6. **Mathematical Consistency:** String theory is remarkably consistent and mathematically robust. It has led to the development of new mathematical tools and concepts, influencing fields like algebraic geometry and topology. The consistency of string theory across various contexts provides strong support for its validity.

7. **Black Hole Physics:** String theory has provided valuable insights into the nature of black holes. It has been instrumental in developing the holographic principle, which suggests that the information contained in a volume of space can be represented by a theory on the boundary of that space. This principle has profound implications for understanding black hole entropy and information paradoxes.

8. **Quantum Gravity:** One of the biggest challenges in theoretical physics is formulating a theory of quantum gravity. String theory offers a candidate for such a theory. By incorporating the graviton, string theory naturally includes a quantum description of gravity, providing a framework to study quantum gravitational effects.

9. **Interconnectedness:** String theory has shown that different versions of the theory are connected through dualities. These dualities imply that

seemingly different physical scenarios are actually equivalent descriptions of the same underlying reality. This interconnectedness hints at a deeper, unified structure of the universe.

10. **Inspirational Power:** Finally, the concept of strings has driven the development of new theoretical ideas and inspired countless researchers. Even if string theory does not turn out to be the final theory of everything, its concepts and mathematical tools have already enriched our understanding of the universe.

Overall, strings provide a versatile and elegant way to address multiple fundamental issues in physics, from resolving infinities in quantum field theory to unifying the forces of nature and offering insights into quantum gravity. The concept of strings extends the possibilities of how we can understand and describe the fundamental constituents of the universe.

History of String Theory

The history of string theory is a fascinating journey of theoretical breakthroughs and evolving ideas, beginning in the late 1960s and continuing to the present day. Here's a detailed look at its development:

1960s: The Birth of String Theory

- **Origins in Particle Physics:** The origins of string theory trace back to attempts to understand the strong nuclear force, which binds protons and neutrons in an atomic nucleus. Physicists Gabriele Veneziano, in 1968, discovered a mathematical formula (the Veneziano amplitude) that described certain properties of strong interactions, sparking interest in a new approach.
- **Early String Models:** Yoichiro Nambu, Holger Bech Nielsen, and Leonard Susskind independently realized in 1970 that Veneziano's formula could be derived from a physical model where particles are represented as one-dimensional strings rather than point-like objects. This was the first conceptualization of string theory.

1970s: The Development and Challenges

- **Bosonic String Theory:** The early formulations of string theory, now known as bosonic string theory, only described bosons, which are particles that mediate forces. However, this theory had significant issues, including requiring 26 dimensions for mathematical consistency and predicting a particle with imaginary mass (a tachyon).
- **Introduction of Supersymmetry:** In the mid-1970s, supersymmetry was introduced to string theory, leading to the development of superstring theory. Supersymmetry posits a symmetry between bosons and fermions (the particles that make up matter). This new approach reduced the

required number of dimensions to 10 and helped address some of the earlier issues.

1980s: The First String Revolution

- **Discovery of Anomaly Cancellation:** In 1984, Michael Green and John Schwarz discovered that anomalies (inconsistencies in the theory) could be cancelled in superstring theory, specifically in ten dimensions. This was a major breakthrough, providing a consistent framework for the theory.
- **Explosion of Interest:** The anomaly cancellation result triggered a surge of interest in string theory. Researchers around the world began exploring its implications, leading to the development of five different consistent superstring theories: Type I, Type IIA, Type IIB, and two versions of heterotic string theory (SO(32) and E8 x E8).

1990s: The Second String Revolution

- **Dualities and M-Theory:** In the mid-1990s, it was realized that the five superstring theories were not distinct but rather connected through various dualities (mathematical equivalences). Edward Witten proposed that these theories were different limits of a more fundamental 11-dimensional theory known as M-theory. This second string revolution unified the different string theories into a single framework.
- **Branes and D-Branes:** The concept of branes (multi-dimensional objects) became central in string theory. D-branes, in particular, were shown to be important in understanding the non-perturbative aspects of the theory, where traditional perturbative methods are insufficient.

2000s: Advancements and Applications

- **Gauge-Gravity Duality:** One of the significant developments was the formulation of the AdS/CFT correspondence by Juan Maldacena in 1997. This duality suggests a relationship between a string theory in a higher-dimensional space (AdS space) and a conformal field theory (CFT) on its boundary. It has provided deep insights into both quantum gravity and strongly interacting gauge theories.
- **String Phenomenology:** Efforts to connect string theory with observable physics, known as string phenomenology, intensified. Researchers aimed to derive the Standard Model of particle physics from string theory, although concrete predictions remain challenging.

Recent Developments

- **Mathematical Advances:** String theory has continued to influence mathematics, particularly in areas like algebraic geometry and number theory. Concepts from string theory have led to new mathematical tools and solved longstanding problems.

- **Quantum Computing and Black Holes:** String theory has also found applications in understanding black holes and quantum computing. The study of black hole entropy and information paradoxes has benefited from string-theoretic insights, particularly through the holographic principle.

Challenges and Criticisms

- **Lack of Experimental Evidence:** A major criticism of string theory is its lack of direct experimental evidence. The predicted strings are incredibly small, making them currently undetectable with existing technology. This has led to debates about the theory's testability.
- **Landscape Problem:** The vast "landscape" of possible string theory solutions, each corresponding to different possible universes with different physical laws, poses a challenge. Identifying the specific vacuum state that matches our universe remains an open problem.

Particles vs. Strings

While particle physics treats particles as point-like entities interacting at specific points, string theory proposes that these particles are different vibrational states of tiny, one-dimensional strings, offering a more unified and potentially comprehensive framework for understanding the fundamental nature of the universe.

1. Fundamental Nature:

- **Particles:** In traditional physics, particles are considered point-like entities with no spatial extent. They are zero-dimensional objects, meaning they are treated as infinitely small points in space. Examples include electrons, quarks, and photons.
- **Strings:** String theory posits that the fundamental constituents of the universe are not point-like particles but one-dimensional "strings." These strings can vibrate at different frequencies, and each vibrational state corresponds to a different particle. Strings have a finite length but are incredibly small, typically on the order of the Planck length (10^{-35} meters).

2. Dimensions:

- **Particles:** In the Standard Model of particle physics, particles exist in a four-dimensional spacetime (three spatial dimensions and one time dimension).
- **Strings:** String theory requires additional spatial dimensions for mathematical consistency. The simplest version of string theory needs ten dimensions (nine spatial and one time), while M-theory, a more comprehensive framework, requires eleven dimensions. These extra

dimensions are compactified, meaning they are curled up so small that they are not observable at human scales.

3. Mathematical Framework:

- **Particles:** The mathematics of particle physics is governed by quantum field theory (QFT). QFT combines quantum mechanics and special relativity to describe how particles interact with each other through fields.
- **Strings:** The mathematics of string theory is much more complex and involves advanced concepts from geometry and topology. String theory uses conformal field theory, supergravity, and other sophisticated mathematical tools to describe the interactions of strings and branes.

4. Interactions:

- **Particles:** In quantum field theory, interactions between particles are described by the exchange of other particles called force carriers (e.g., photons for electromagnetism, gluons for the strong force). These interactions occur at points where the particles meet.
- **Strings:** In string theory, interactions occur when strings split and join. Because strings are extended objects, these interactions are spread out over a small but finite area, which helps avoid certain mathematical infinities that plague point-particle theories.

5. Force Unification:

- **Particles:** The Standard Model successfully unifies three of the four fundamental forces (electromagnetism, weak nuclear force, and strong nuclear force) but does not include gravity. General relativity, which describes gravity, remains separate and is not easily reconcilable with quantum mechanics.
- **Strings:** String theory naturally includes gravity, as one of the vibrational modes of a string corresponds to the graviton, the hypothetical quantum particle that mediates gravitational interactions. This unification of gravity with the other forces is one of string theory's major achievements.

6. Supersymmetry:

- **Particles:** The Standard Model does not inherently include supersymmetry (SUSY), though SUSY can be added as an extension. SUSY proposes that each particle has a superpartner with a different spin.
- **Strings:** Supersymmetry is a crucial component of superstring theory. It helps cancel out certain mathematical inconsistencies and allows for a more coherent theoretical framework. Superstring theory requires supersymmetry to work correctly.

7. Phenomenology:

- **Particles:** The predictions of the Standard Model have been confirmed through numerous experiments. It accurately describes the behavior of known particles and forces at accessible energy scales.
- **Strings:** Direct experimental evidence for string theory is still lacking. The strings are so small that detecting them with current technology is not possible. However, string theory has provided valuable insights and frameworks for understanding high-energy physics and quantum gravity.

8. Conceptual Paradigm:

- **Particles:** The point-particle paradigm treats particles as the most fundamental units, each with specific properties like mass, charge, and spin.
- **Strings:** The string paradigm views particles as different manifestations of one fundamental entity—a vibrating string. The properties of particles arise from the different vibrational modes of these strings.

9. Predictive Power:

- **Particles:** The Standard Model makes precise predictions that have been confirmed experimentally, such as the existence of the Higgs boson.
- **Strings:** String theory has a vast landscape of possible solutions, leading to many possible universes with different physical laws. This makes it challenging to make specific, testable predictions about our universe.

10. Theoretical Developments:

- **Particles:** Quantum field theory and the Standard Model are well-developed and have a solid experimental foundation. Research continues to explore physics beyond the Standard Model, such as dark matter and neutrino masses.
- **Strings:** String theory is still developing, with ongoing research into understanding its full implications, including M-theory, brane dynamics, and the holographic principle. It remains a promising but speculative framework for a unified theory of fundamental forces.

String Theory in a Nutshell

String theory is a theoretical framework in which the fundamental particles of nature are not point-like objects but one-dimensional "strings." These strings can vibrate at different frequencies, and the various vibrational modes correspond to different particles, such as electrons, quarks, and photons. Here's a concise overview of string theory:

1. Fundamental Idea:

- **Strings as Fundamental Entities:** Unlike traditional particle physics, · which considers particles as point-like, string theory proposes that the basic building blocks of the universe are tiny, vibrating strings. These strings can be open (with two endpoints) or closed (forming loops).

2. Dimensions:

- **Extra Dimensions:** For string theory to be mathematically consistent, it requires more than the familiar four dimensions (three of space and one of time). The simplest versions of string theory require ten dimensions (nine spatial and one time), while M-theory, an extension, requires eleven dimensions. These additional dimensions are compactified, meaning they are curled up so small that they are not observable at everyday scales.

3. Unification of Forces:

- **Inclusion of Gravity:** One of the major achievements of string theory is its natural inclusion of gravity. It predicts the existence of a particle called the graviton, which mediates the gravitational force, thereby providing a framework for unifying all fundamental forces (gravity, electromagnetism, weak nuclear force, and strong nuclear force) into a single theory.

4. Supersymmetry:

- **Symmetry Between Particles:** String theory incorporates supersymmetry, a theoretical symmetry that pairs each particle with a "superpartner" that differs in spin. Supersymmetry helps to resolve various theoretical issues, such as the cancellation of certain infinities.

5. Vibrational Modes:

- **Different Particles:** The different particles observed in nature are understood as different vibrational modes of strings. Just as different notes from a musical string depend on how it vibrates, different particles arise from the various vibrational patterns of fundamental strings.

6. Mathematical Framework:

- **Complex Mathematics:** The mathematics of string theory involves advanced concepts from geometry, topology, and conformal field theory. This complexity allows for a rich and detailed description of particle interactions and the fundamental structure of the universe.

7. Dualities and M-Theory:

- **Interconnected Theories:** String theory is not a single theory but a network of five related theories connected by dualities (mathematical equivalences). These dualities reveal that different string theories are just different perspectives on the same underlying physics. M-theory is a proposed unified theory that encompasses all five string theories and suggests an 11-dimensional framework.

8. Challenges and Criticisms:

- **Lack of Experimental Evidence:** One of the main criticisms of string theory is the absence of direct experimental evidence. The strings are so small that detecting them with current technology is not possible. However, string theory has provided profound insights into theoretical physics and mathematics.
- **Landscape Problem:** String theory has a vast number of possible solutions, each corresponding to a different set of physical laws. This "landscape" of solutions makes it difficult to identify which, if any, describes our universe.

9. Applications and Insights:

- **Black Holes and Holography:** String theory has offered valuable insights into the nature of black holes and the structure of spacetime. The holographic principle, derived from string theory, suggests that the information contained in a volume of space can be described by a theory on its boundary, providing a new perspective on quantum gravity.

10. Ongoing Research:

- **Continued Exploration:** Researchers continue to explore the implications of string theory, seeking to connect it more closely with observable phenomena and to resolve outstanding theoretical challenges. The theory remains a leading candidate for a unified theory of fundamental forces and continues to inspire new directions in theoretical physics.

In a nutshell, string theory offers a bold and innovative framework for understanding the fundamental structure of the universe, positing that all particles are manifestations of tiny, vibrating strings. Its mathematical beauty and potential to unify all fundamental forces make it a captivating area of ongoing research in theoretical physics.

Dimensions Beyond the Observable Universe

String theory introduces the concept of extra dimensions beyond the familiar three spatial dimensions and one-time dimension. These additional dimensions are crucial for the mathematical consistency of the theory and provide a richer framework for understanding the universe.

In our everyday experience, we perceive only three dimensions of space: length, width, and height. Along with the dimension of time, this makes up the four-dimensional spacetime of our observable universe. However, string theory posits that there are more dimensions than we can directly observe. The simplest versions of string theory require ten dimensions: nine of space and one of time. In some versions, like M-theory, there are eleven dimensions, which include ten spatial dimensions and one-time dimension.

Why do we need these extra dimensions? The additional dimensions in string theory allow for the necessary mathematical structures to describe all fundamental forces, including gravity, in a unified manner. Without these extra dimensions, the equations of string theory do not work correctly.

These extra dimensions are not visible in our daily lives because they are compactified. Compactification is a process where extra dimensions are curled up or folded into extremely small sizes, typically at the scale of the Planck length (10^{-35} meters). This size is so small that it is beyond the reach of current experimental detection.

Imagine a garden hose lying on the ground. From a distance, it looks like a one-dimensional line. However, if you zoom in close enough, you can see it has a second dimension as well: the circular cross-section of the hose. Similarly, the extra dimensions in string theory are compactified into shapes so tiny that we do not notice them in our everyday experience.

There are various ways to compactify these extra dimensions. The shapes of these compactified dimensions are often described by complex geometries known as Calabi-Yau manifolds. These manifolds are intricate, multi-dimensional shapes that satisfy certain mathematical properties, making them suitable for the extra dimensions in string theory. The specific geometry of the compactified dimensions can influence the physical properties of our universe, such as the types of particles and forces we observe.

In addition to compactification, there are different ways to consider the extra dimensions. One approach involves large extra dimensions, which are still small but significantly larger than the Planck length. This idea suggests that while we cannot see these dimensions directly, their effects might be detectable at high energy scales, such as those probed by particle colliders like the Large Hadron Collider (LHC).

Another intriguing concept in string theory is the idea of branes, which are multi-dimensional objects that exist within these higher-dimensional spaces. Our observable universe could be a three-dimensional brane embedded in a higher-dimensional space. In this scenario, while we are confined to our three-dimensional brane, gravity and other forces might propagate through the higher-dimensional space. This could explain why gravity is much weaker compared to the other fundamental forces; it is diluted as it spreads out into the extra dimensions.

The Role of Mathematics in String Theory

The complex and abstract nature of string theory requires sophisticated mathematical tools and concepts to describe the behavior of strings and the extra dimensions they inhabit. Here's a look at the role of mathematics in string theory:

1. Foundational Framework

- **Equations of Motion:** The behavior of strings is governed by equations derived from the principles of quantum mechanics and special relativity. These equations are more intricate than those for point particles, involving partial differential equations that describe the dynamics of one-dimensional objects.
- **Action Principle:** The Polyakov action, a generalization of the action principle used in classical mechanics, is fundamental in string theory. This action is expressed using advanced mathematical constructs like conformal field theory, which describes how strings propagate through spacetime.

2. Extra Dimensions and Geometry

- **Compactification:** The concept of extra dimensions in string theory requires compactification, where additional dimensions are curled up into small, complex shapes. The mathematics of compactification involves intricate geometries such as Calabi-Yau manifolds. These manifolds have specific properties that allow for the consistent behavior of strings and preserve supersymmetry.
- **Topology:** The study of topology, which deals with properties of space that are preserved under continuous deformations, is crucial. Different topological structures can lead to different physical phenomena in the compactified dimensions, influencing particle properties and interactions.

3. Symmetries and Dualities

- **Supersymmetry:** String theory incorporates supersymmetry, a mathematical symmetry that pairs bosons (force-carrying particles) with fermions (matter particles). This symmetry helps solve various theoretical problems, such as the hierarchy problem and the unification of forces.
- **Dualities:** Dualities are mathematical equivalences between different string theories, revealing that they are just different descriptions of the same underlying physics. These dualities, such as S-duality and T-duality, involve

transformations that link strong and weak coupling regimes and large and small distance scales, respectively.

4. Advanced Mathematical Concepts

* **Conformal Field Theory (CFT):** CFT is used to describe the quantum behavior of strings. It involves the study of fields that are invariant under conformal transformations, which preserve angles but not distances. This is essential for understanding how strings interact and propagate.
* **Algebraic Geometry:** This branch of mathematics studies solutions to systems of polynomial equations. It is used to analyze the shapes and structures of the compactified dimensions and their impact on string dynamics.
* **Moduli Spaces:** Moduli spaces represent the set of all possible shapes and sizes of the compactified dimensions. They are essential for understanding the landscape of string theory solutions and the parameters that define different possible universes.

5. Holography and Black Hole Physics

* **AdS/CFT Correspondence:** The AdS/CFT correspondence, formulated by Juan Maldacena, is a duality between a string theory in anti-de Sitter (AdS) space and a conformal field theory (CFT) on its boundary. This mathematical relationship provides deep insights into quantum gravity and black hole physics, showing how information in a higher-dimensional space can be encoded on a lower-dimensional boundary.
* **Black Hole Entropy:** String theory provides a microscopic explanation for the entropy of black holes, as described by the Bekenstein-Hawking formula. The mathematics of string theory allows for the counting of microstates that contribute to the black hole's entropy, offering a statistical mechanics interpretation of gravitational phenomena.

6. Mathematical Rigor and Predictions

* **Consistency Checks:** The mathematical consistency of string theory is verified through rigorous calculations and checks, ensuring that anomalies are canceled, and equations are coherent. This rigorous approach has led to numerous breakthroughs and validations of the theory's internal consistency.
* **Predictions and Phenomenology:** Mathematics is used to derive predictions from string theory that can, in principle, be tested experimentally. This involves calculating the properties of particles, forces, and possible deviations from known physics that might be observable in high-energy experiments or cosmological observations.

7. Influence on Mathematics

- **Mathematical Innovation:** String theory has led to new developments in mathematics, inspiring advances in areas such as mirror symmetry, knot theory, and even number theory. The interplay between string theory and mathematics is a two-way street, with each field enriching the other.

String Theory and Quantum Mechanics

String theory and quantum mechanics are deeply intertwined, with string theory offering a framework that incorporates and extends the principles of quantum mechanics to describe the fundamental nature of the universe. Here's how string theory relates to and builds upon quantum mechanics.

Quantum Mechanics: The Foundation

Quantum mechanics is the branch of physics that describes the behavior of particles at the smallest scales. It introduces the idea that particles, such as electrons and photons, exhibit both wave-like and particle-like properties. The key principles include wave-particle duality, quantization, and uncertainty.

1. **Wave-Particle Duality:** Particles can behave like waves, with properties such as interference and diffraction.
2. **Quantization:** Certain properties, like energy levels in an atom, can only take on discrete values.
3. **Uncertainty Principle:** It is impossible to simultaneously know both the exact position and momentum of a particle.

Extending Quantum Mechanics with Strings

In traditional quantum mechanics, particles are point-like. String theory changes this by proposing that the fundamental entities are not points but one-dimensional strings. These strings can vibrate at different frequencies, and each vibrational mode corresponds to a different particle. This shift from points to strings offers several advantages and deeper insights.

The Role of Strings

1. **Vibrational Modes:** In string theory, the different particles observed in nature are understood as different vibrational states of strings. Just as a guitar string can produce different notes depending on how it vibrates, strings can produce different particles, including those that mediate forces like photons and gluons.
2. **Avoiding Singularities:** Point-like particles can lead to mathematical singularities, where physical quantities become infinite. Strings, being extended objects, spread out interactions over a finite area, helping to avoid these problematic infinities.

Quantum Field Theory and Strings

Quantum field theory (QFT) is the framework that combines quantum mechanics and special relativity to describe the interactions of particles through fields. String theory can be seen as a natural extension of QFT, where instead of fields interacting at points, strings interact by joining and splitting.

1. **String Interactions:** When two strings collide, they do not interact at a single point. Instead, they merge and split over a region of space, which softens the interaction and avoids the infinities that plague point-particle interactions in QFT.
2. **Graviton Emergence:** In string theory, the graviton, the hypothetical particle that mediates the force of gravity, naturally arises as one of the vibrational modes of a closed string. This is a significant step towards unifying gravity with the other fundamental forces within a quantum framework.

Supersymmetry and Quantum Mechanics

String theory incorporates supersymmetry, a theoretical symmetry that pairs each particle with a superpartner. Supersymmetry is crucial for solving various theoretical problems in quantum mechanics and field theory.

1. **Cancellation of Infinities:** Supersymmetry helps cancel out certain infinities that arise in quantum field theories. These cancellations are necessary for the mathematical consistency of string theory.
2. **Unification of Forces:** Supersymmetry facilitates the unification of the electromagnetic, weak, and strong nuclear forces with gravity. This unification is a major goal of theoretical physics.

Quantum Mechanics in Extra Dimensions

String theory requires extra spatial dimensions beyond the familiar three. These extra dimensions are compactified, meaning they are curled up into tiny shapes that are not directly observable.

1. **Calabi-Yau Manifolds:** The shapes of these extra dimensions are often described by complex geometries known as Calabi-Yau manifolds. The properties of these compactified dimensions affect the vibrational modes of strings, and thus the physical properties of particles.
2. **Higher-Dimensional Quantum Mechanics:** The quantum mechanics of strings in these higher-dimensional spaces is more complex but provides a richer structure for understanding fundamental interactions.

Quantum Fluctuations and String Theory

Quantum mechanics introduces the concept of quantum fluctuations, where particles can temporarily appear and disappear. String theory incorporates these fluctuations but extends them to strings.

1. **String Fluctuations:** Strings can vibrate and fluctuate in higher-dimensional space, leading to a diverse spectrum of particles and interactions.
2. **Virtual Strings:** Just as virtual particles exist in quantum field theory, virtual strings exist in string theory, contributing to the interactions between physical strings.

Mathematical Rigor

The mathematics of string theory is intricate, involving advanced concepts from geometry, topology, and algebra. This mathematical framework is essential for describing the quantum behavior of strings and ensuring the theory's consistency.

1. **Conformal Field Theory:** This is used to describe the quantum states of strings, ensuring that the theory respects the principles of quantum mechanics.
2. **Dualities:** Mathematical dualities in string theory reveal deep connections between seemingly different physical scenarios, reflecting the underlying unity of quantum phenomena.

String Theory vs. Classical Physics

String theory and classical physics offer two very different frameworks for understanding the universe. Classical physics, based on the laws formulated by Isaac Newton and later refined by others, describes the macroscopic world we experience daily. In contrast, string theory is a modern framework aimed at explaining the fundamental nature of all particles and forces at the smallest scales. Here's a detailed comparison of string theory and classical physics:

1. Fundamental Entities

- **Classical Physics:** In classical physics, the universe is composed of point-like particles. These particles have well-defined positions and velocities at any given time.
- **String Theory:** In string theory, the fundamental entities are one-dimensional strings. These strings can vibrate at different frequencies, and their vibrational modes correspond to different particles. Unlike point particles, strings have a finite size, typically on the order of the Planck length (10^{-35} meters).

2. Dimensions

- **Classical Physics:** Classical physics operates in a three-dimensional space with one dimension of time. Newtonian mechanics and classical electromagnetism work within this familiar 3+1 dimensional framework.
- **String Theory:** String theory requires additional spatial dimensions for mathematical consistency. The simplest versions require ten dimensions (nine spatial and one time), while M-theory, an extension, requires eleven dimensions. These extra dimensions are compactified, meaning they are curled up into tiny, unobservable shapes.

3. Nature of Forces

- **Classical Physics:** Forces in classical physics are described by fields and equations of motion. Newton's laws describe gravitational force, while Maxwell's equations describe electromagnetism. These laws operate on macroscopic scales and treat forces as acting instantaneously at a distance.
- **String Theory:** In string theory, forces arise from the interactions of strings. The graviton, a hypothetical quantum particle, mediates gravity and is a vibrational mode of a closed string. Other forces, like electromagnetism, are also explained through different string vibrations. String theory naturally unifies all fundamental forces, including gravity, electromagnetism, and the nuclear forces, within a single framework.

4. Mathematical Framework

- **Classical Physics:** The mathematics of classical physics involves calculus, differential equations, and vector analysis. It deals with deterministic equations of motion, where knowing initial conditions allows for precise predictions of future states.
- **String Theory:** The mathematics of string theory is far more complex, involving advanced concepts from algebraic geometry, topology, and conformal field theory. String theory uses these sophisticated tools to describe the quantum behavior of strings and the geometry of extra dimensions.

5. Determinism vs. Probability

- **Classical Physics:** Classical physics is deterministic. Given the initial conditions of a system, its future behavior can be predicted exactly. This is epitomized by Newton's laws of motion.
- **String Theory:** String theory, like all quantum theories, is probabilistic. It incorporates the principles of quantum mechanics, where outcomes are described by probabilities rather than certainties. The behavior of strings is governed by quantum mechanics, leading to inherent uncertainties.

6. Scale of Applicability

- **Classical Physics:** Classical physics excels at describing the behavior of macroscopic objects, from planets and stars to everyday objects like cars and buildings. It fails at atomic and subatomic scales, where quantum effects become significant.

- **String Theory:** String theory aims to describe the fundamental nature of the universe at the smallest scales, including the Planck scale, where quantum gravitational effects become important. It encompasses and extends quantum mechanics and general relativity.

7. Unification of Forces

- **Classical Physics:** In classical physics, different forces are described by separate theories. Newtonian mechanics describes gravity, while Maxwell's equations handle electromagnetism. There is no unification of forces.
- **String Theory:** String theory provides a unified framework where all forces, including gravity, emerge from the interactions of strings. This unification is a major goal of theoretical physics, seeking to reconcile quantum mechanics with general relativity.

8. Concept of Space and Time

- **Classical Physics:** In classical physics, space and time are absolute and independent. Newton's laws operate within this fixed backdrop, and time flows uniformly everywhere.
- **String Theory:** String theory incorporates the ideas of general relativity, where space and time are dynamic and can be curved by the presence of mass and energy. The extra dimensions of string theory add complexity to our understanding of spacetime.

9. Experimental Evidence

- **Classical Physics:** Classical physics is well-supported by experimental evidence. Its predictions match observations in the macroscopic world with high precision.
- **String Theory:** Direct experimental evidence for string theory is still lacking. The strings are too small to be detected with current technology. However, string theory has provided valuable insights and frameworks for understanding high-energy physics and quantum gravity.

10. Conceptual Paradigms

- **Classical Physics:** Classical physics relies on straightforward, intuitive concepts like forces, motion, and energy. It operates within a well-defined framework that aligns closely with our everyday experiences.
- **String Theory:** String theory introduces abstract and counterintuitive concepts, such as additional dimensions and vibrating strings. It requires a shift in perspective to understand the fundamental nature of particles and forces.

In short, classical physics provides a deterministic, intuitive framework for understanding macroscopic phenomena, while string theory offers a probabilistic, abstract framework aimed at unifying all fundamental forces at the smallest scales.

Future of String Theory

The future of string theory holds many possibilities and challenges, as it continues to be a central topic in theoretical physics. Here's a detailed look at what the future may hold for string theory:

1. Experimental Validation

- **Advancements in Technology:** One of the biggest challenges for string theory is the lack of direct experimental evidence. Future advancements in technology, such as more powerful particle accelerators or new methods for detecting high-energy phenomena, could provide the means to test predictions of string theory. Detecting supersymmetric particles or other indirect evidence could lend support to the theory.
- **Cosmological Observations:** Observations of the early universe, such as those from the cosmic microwave background (CMB) or gravitational waves, could provide clues about the validity of string theory. These observations might reveal signatures consistent with extra dimensions or other string-theoretic predictions.

2. Mathematical Developments

- **Refinement of Theories:** Continued mathematical research will refine and expand string theory. New mathematical tools and techniques could provide deeper insights into the theory's structure and resolve some of its current ambiguities. Advancements in areas such as algebraic geometry, topology, and quantum field theory will be important.
- **Computational Techniques:** Increased computational power and sophisticated algorithms will enable more detailed simulations and models. This could help physicists explore the vast landscape of string theory solutions and better understand the properties of different compactified dimensions.

3. Unification and Theory of Everything

- **Integration with Other Theories:** String theory aims to be a "theory of everything" by unifying all fundamental forces, including gravity. Future work will focus on integrating string theory with other approaches to quantum gravity, such as loop quantum gravity, to develop a more comprehensive framework.
- **Resolution of the Landscape Problem:** The string theory landscape, with its vast number of possible solutions, poses a significant challenge. Future research may find ways to narrow down this landscape to a smaller

set of viable solutions that correspond to our universe, potentially through new principles or selection mechanisms.

4. Practical Applications

- **New Physics Beyond the Standard Model:** String theory could lead to discoveries of new physics beyond the Standard Model. This might include new particles, forces, or interactions that could be experimentally tested and applied in various technologies.
- **Insights into Quantum Computing:** Concepts from string theory, such as brane dynamics and holography, might provide new insights into quantum computing and information theory. These could lead to advancements in computational methods and technologies.

5. Cross-Disciplinary Impact

- **Influence on Mathematics:** String theory has already had a profound impact on mathematics, leading to new discoveries and methods. This cross-disciplinary influence will likely continue, with string theory inspiring new directions in mathematical research.
- **Impact on Other Fields:** Insights from string theory could influence other areas of science, such as cosmology, condensed matter physics, and even biology. The conceptual frameworks developed in string theory might provide new ways to approach problems in these fields.

6. Educational and Collaborative Efforts

- **Training the Next Generation:** Continued emphasis on education and training in string theory and related fields will be crucial. Developing comprehensive curricula and fostering collaboration among institutions will help cultivate the next generation of theoretical physicists.
- **International Collaboration:** Global collaboration among physicists will be essential to advance string theory. Large-scale projects, conferences, and collaborative research efforts will play a key role in sharing knowledge and making progress.

7. Philosophical and Conceptual Developments

- **Foundations of Physics:** String theory challenges and extends our understanding of the foundations of physics. Future work will continue to explore the philosophical implications of concepts like extra dimensions, the multiverse, and the nature of reality.
- **Interpretations of Quantum Mechanics:** String theory's integration with quantum mechanics may provide new insights into the interpretation of quantum phenomena, potentially resolving long-standing debates and paradoxes.

8. Challenges and Criticisms

- **Addressing Criticisms:** String theory faces criticisms for its lack of empirical evidence and the vast number of possible solutions. Addressing these criticisms through theoretical, experimental, and philosophical work will be crucial for its future development.
- **Balancing Speculation and Rigor:** Striking a balance between speculative ideas and rigorous, testable predictions will be essential. Ensuring that string theory remains grounded in empirical science while exploring bold new concepts will guide its progress.

The future of string theory is filled with promise and complexity. As physicists and mathematicians continue to develop the theory, the potential for new discoveries and deeper understanding of the universe remains high. Whether through technological advancements, mathematical breakthroughs, or novel experimental techniques, string theory is likely to remain at the forefront of theoretical physics, pushing the boundaries of our knowledge and offering insights into the nature of reality.

CHAPTER 2: THE BIRTH OF STRING THEORY

Early Theories and Discoveries

String theory began its journey in the late 1960s, emerging from the quest to understand the strong nuclear force, which binds protons and neutrons in the nucleus. The earliest steps toward string theory were marked by a desire to describe this force in a way that existing models couldn't.

Veneziano's Insight

In 1968, Gabriele Veneziano, an Italian physicist, made a groundbreaking discovery. He found a mathematical formula, now known as the Veneziano amplitude, that described the scattering of particles involved in the strong interaction. This formula matched experimental data remarkably well but lacked a clear physical interpretation. Veneziano's formula hinted at a deeper underlying structure, paving the way for the development of string theory.

The Birth of String Theory

The next significant step came when Yoichiro Nambu, Holger Bech Nielsen, and Leonard Susskind independently realized that Veneziano's amplitude could be explained by a model where particles are not point-like but are instead one-dimensional objects—strings. This insight marked the birth of string theory. They proposed that the particles we observe are actually different vibrational modes of these fundamental strings.

Early Models: Bosonic String Theory

The earliest versions of string theory are now known as bosonic string theory. These models describe only bosons, which are particles that mediate forces, like photons and gluons. Bosonic string theory required 26 dimensions for mathematical consistency, far beyond the familiar four dimensions of spacetime. Moreover, it predicted the existence of a particle with imaginary mass, a tachyon, which suggested instability in the theory.

The Introduction of Supersymmetry

In the mid-1970s, physicists began to incorporate the idea of supersymmetry into string theory. Supersymmetry proposes that each particle has a superpartner with a spin differing by half a unit. This inclusion led to the development of superstring theory, which reduced the required number of dimensions from 26 to 10. Supersymmetry also helped eliminate the problematic tachyon and brought stability to the theory.

Five Consistent Superstring Theories

By the mid-1980s, researchers had identified five different consistent superstring theories: Type I, Type IIA, Type IIB, and two versions of heterotic string theory (SO(32) and E8 x E8). Each of these theories described different types of strings and interactions but shared common foundational principles. The diversity of these theories initially seemed like a problem, as physicists hoped for a single unified theory.

Anomaly Cancellation

A major breakthrough came in 1984 when Michael Green and John Schwarz demonstrated that anomalies in superstring theory could be cancelled out in ten dimensions. This result showed that the theory could be free of internal inconsistencies, leading to a surge of interest and research in the field. The demonstration of anomaly cancellation provided a solid foundation for the further development of string theory.

Dualities and Unification

In the 1990s, physicists discovered that the five different superstring theories were not separate after all. They were connected through various dualities—mathematical transformations that showed these theories were just different aspects of the same underlying framework. Edward Witten proposed that these theories were different limits of an even more fundamental theory known as M-theory, which required eleven dimensions. This realization, known as the second string revolution, unified the different string theories into a single cohesive framework.

Branes and Higher Dimensions

The concept of branes (short for membranes) emerged as a crucial element of string theory. Branes are multi-dimensional objects within the theory. For example, a one-dimensional brane is a string, while a two-dimensional brane is a membrane. These branes can exist in various dimensions and interact with strings in complex ways, adding richness to the theory's structure.

Influence on Mathematics and Physics

The development of string theory has profoundly influenced both mathematics and theoretical physics. It has introduced new mathematical concepts and techniques, solving problems and opening new areas of study. String theory has also provided insights into black holes and the nature of spacetime, influencing our understanding of quantum gravity.

Ongoing Research and Challenges

Despite its successes, string theory faces challenges, particularly its lack of direct experimental evidence. The theory's predictions often involve energy scales far beyond current experimental capabilities. Researchers continue to explore ways to connect string theory more directly with observable phenomena and to address the vast landscape of possible solutions within the theory.

Key Scientists and Contributors

String theory has been shaped and advanced by many brilliant scientists since its inception. Here are some of the key contributors and their significant contributions to the development of string theory:

Gabriele Veneziano

- **Contribution:** In 1968, Veneziano discovered a mathematical formula, the Veneziano amplitude, that described the scattering of particles in strong interactions. This formula laid the groundwork for the development of string theory.
- **Significance:** His work provided a successful model for particle interactions and suggested that there was a deeper structure to these interactions, which eventually led to the concept of strings.

Yoichiro Nambu

- **Contribution:** Nambu was one of the first to realize that Veneziano's amplitude could be derived from a physical model where particles are one-dimensional strings rather than point-like objects.
- **Significance:** His insights helped establish the foundational idea that particles could be understood as different vibrational modes of strings.

Leonard Susskind

- **Contribution:** Susskind independently proposed that the Veneziano amplitude could be explained by string theory, reinforcing the idea that particles are vibrating strings.
- **Significance:** He was instrumental in popularizing the concept of strings in particle physics and helped solidify the theoretical underpinnings of string theory.

Holger Bech Nielsen

- **Contribution:** Alongside Nambu and Susskind, Nielsen recognized that the Veneziano amplitude could be explained through string theory. He contributed to the early development and promotion of the theory.
- **Significance:** His work helped in establishing string theory as a viable framework for understanding particle interactions.

Michael Green

- **Contribution:** Green, along with John Schwarz, made a significant breakthrough in 1984 by demonstrating that anomalies in superstring theory could be cancelled out, ensuring the theory's mathematical consistency.
- **Significance:** This anomaly cancellation result, known as the Green-Schwarz mechanism, provided strong evidence for the validity of superstring theory and sparked renewed interest in the field.

John Schwarz

- **Contribution:** Schwarz worked closely with Michael Green to show that superstring theory is anomaly-free in ten dimensions.
- **Significance:** Their work laid the foundation for modern string theory and helped establish it as a leading candidate for a theory of everything.

Edward Witten

- **Contribution:** Witten made numerous contributions to string theory, including proposing M-theory, which unifies the five different superstring theories. He also introduced important concepts like branes and dualities.
- **Significance:** Witten's work has been pivotal in advancing string theory, providing deeper insights into its mathematical structure and unifying its various formulations.

David Gross

- **Contribution:** Gross, along with Frank Wilczek and David Politzer, developed the theory of asymptotic freedom in quantum chromodynamics, which indirectly influenced string theory by advancing the understanding of fundamental forces.
- **Significance:** His work in particle physics has helped shape the broader context in which string theory developed.

Frank Wilczek

- **Contribution:** Wilczek's work on asymptotic freedom and his contributions to quantum field theory have had a significant impact on theoretical physics.
- **Significance:** His research has influenced the development of theories that complement and intersect with string theory.

David Politzer

- **Contribution:** Politzer, alongside Gross and Wilczek, developed the concept of asymptotic freedom, which describes how the strong force behaves at different energy scales.
- **Significance:** His contributions to understanding fundamental forces have provided a foundation upon which string theory builds.

Juan Maldacena

- **Contribution:** Maldacena formulated the AdS/CFT correspondence, a conjecture that relates a string theory in a higher-dimensional space to a lower-dimensional conformal field theory.
- **Significance:** This duality has provided profound insights into quantum gravity and has been a major development in theoretical physics, influencing a wide range of research in string theory.

Cumrun Vafa

- **Contribution:** Vafa has made significant contributions to the understanding of string theory, particularly in the areas of compactification and the role of extra dimensions.
- **Significance:** His work has helped to elucidate the structure of string theory and its implications for particle physics and cosmology.

Joe Polchinski

- **Contribution:** Polchinski is known for his work on D-branes, which are essential objects in string theory that can exist in various dimensions.
- **Significance:** His contributions have been crucial in understanding the dynamics of strings and branes, which are fundamental components of string theory.

Lisa Randall

- **Contribution:** Randall has worked on the implications of extra dimensions and has proposed models that explore how these dimensions might be hidden from our everyday experience.
- **Significance:** Her research has provided new perspectives on how string theory and extra dimensions might manifest in the observable universe.

Brian Greene

- **Contribution:** Greene has been a prominent figure in communicating the complexities of string theory to the public through his books and lectures.
- **Significance:** His work has helped to popularize string theory and make it accessible to a broader audience, inspiring interest and understanding in the field.

These scientists, along with many others, have been instrumental in the development and advancement of string theory. Their collective contributions have shaped string theory into a comprehensive framework that seeks to unify all fundamental forces and particles.

Evolution from Particle Physics

The journey from particle physics to string theory began with the quest to understand the strong nuclear force. In the 1960s, physicists were grappling with the complexities of the strong force, which binds protons and neutrons in an atomic nucleus. The existing models couldn't fully explain the behavior of particles involved in this interaction.

Early Challenges in Particle Physics

Physicists used quantum field theory to describe particle interactions. Quantum electrodynamics (QED) successfully explained electromagnetism, and quantum chromodynamics (QCD) described the strong force. However, QCD faced difficulties at high energy scales. The equations led to infinities, making the calculations problematic. Physicists needed a new approach to tame these infinities and provide a more comprehensive understanding of particle interactions.

Veneziano's Breakthrough

In 1968, Gabriele Veneziano made a crucial discovery. He found a mathematical formula, the Veneziano amplitude, which accurately described the scattering of particles in the strong interaction. This formula matched experimental data but lacked a clear physical interpretation. It hinted at a deeper structure underlying particle interactions, suggesting that particles might not be point-like but have some extended structure.

Emergence of String Theory

Physicists Yoichiro Nambu, Holger Bech Nielsen, and Leonard Susskind independently realized that Veneziano's amplitude could be derived from a model where particles are not points but one-dimensional objects—strings. This was the birth of string theory. They proposed that particles we observe are different vibrational modes of these fundamental strings. The strings could vibrate in various ways, with each mode corresponding to a different particle.

Bosonic String Theory

The initial versions of string theory, known as bosonic string theory, described only bosons, the particles that mediate forces. This theory required 26 dimensions for mathematical consistency, far more than the familiar four dimensions of spacetime. Additionally, it predicted the existence of a particle with imaginary mass, called a tachyon, indicating instability in the theory. Despite these issues, bosonic string theory laid the groundwork for further developments.

Introduction of Supersymmetry

In the 1970s, supersymmetry was introduced to string theory. Supersymmetry proposed that each particle has a superpartner with a spin differing by half a unit. Incorporating supersymmetry into string theory led to the development of superstring theory. This new theory required only ten dimensions instead of 26 and eliminated the problematic tachyon. Supersymmetry also helped stabilize the theory and make it more consistent with observed particle physics.

The Five Superstring Theories

By the mid-1980s, physicists had developed five different consistent superstring theories: Type I, Type IIA, Type IIB, and two versions of heterotic string theory (SO(32) and E8 x E8). Each theory described different types of strings and interactions but shared common foundational principles. Initially, the existence of five distinct theories seemed problematic for the goal of a single unified theory.

Anomaly Cancellation and Validation

In 1984, Michael Green and John Schwarz made a breakthrough by demonstrating that anomalies in superstring theory could be cancelled out, ensuring the theory's mathematical consistency. This result, known as the Green-Schwarz mechanism, provided strong evidence for the validity of superstring theory. It sparked a surge of interest and research in string theory, establishing it as a leading candidate for a unified theory of fundamental forces.

Unification through Dualities

In the 1990s, physicists discovered that the five superstring theories were not separate after all. They were connected through various dualities—mathematical transformations showing these theories as different aspects of the same underlying framework. Edward Witten proposed that these theories were different limits of an even more fundamental theory known as M-theory, which required eleven dimensions. This realization, known as the second string revolution, unified the different string theories into a single cohesive framework.

Branes and Higher Dimensions

The concept of branes emerged as a crucial element of string theory. Branes are multi-dimensional objects within the theory. For example, a one-dimensional brane is a string, while a two-dimensional brane is a membrane. These branes can exist in various dimensions and interact with strings in complex ways, adding richness to the theory's structure.

Impact on Mathematics and Physics

The evolution from particle physics to string theory has profoundly influenced both fields. String theory has introduced new mathematical concepts and techniques, solving problems and opening new areas of study. It has also provided insights into black holes and the nature of spacetime, influencing our understanding of quantum gravity.

The First String Revolution

The first string revolution began in the early 1980s and transformed the landscape of theoretical physics. Before this period, string theory was a relatively obscure idea. It had promising concepts but faced significant challenges, including mathematical inconsistencies and the lack of a clear connection to observable phenomena.

Anomaly Cancellation

The breakthrough that ignited the first string revolution came in 1984, when Michael Green and John Schwarz demonstrated that superstring theory could be free of anomalies. Anomalies are mathematical inconsistencies that can render a theory invalid. Green and Schwarz showed that, in ten dimensions, the anomalies in superstring theory cancelled out perfectly. This result, known as the Green-Schwarz mechanism, provided the first solid evidence that string theory could be a consistent framework for describing fundamental particles and forces.

Superstring Theory's Appeal

Superstring theory's ability to include gravity naturally was one of its most appealing features. Unlike previous theories that struggled to incorporate gravity with quantum mechanics, string theory predicted a particle called the graviton, which mediates the gravitational force. This prediction positioned string theory as a leading candidate for a unified theory of everything, aiming to describe all fundamental forces—gravity, electromagnetism, and the strong and weak nuclear forces—within a single framework.

Five Consistent Theories

By the mid-1980s, researchers identified five distinct but consistent superstring theories: Type I, Type IIA, Type IIB, and two versions of heterotic string theory (SO(32) and E8 x E8). Each theory described different types of strings and interactions but shared underlying principles. The existence of five different theories initially seemed like a setback, as physicists hoped for a single, unique theory.

Heterotic String Theory

The heterotic string theories combined elements of both bosonic and superstring theories. This hybrid approach allowed for a richer and more flexible framework, accommodating various symmetry groups and particle interactions. The E8 x E8 version of heterotic string theory, in particular, showed promise for describing the Standard Model of particle physics, which was a significant step towards connecting string theory with observable phenomena.

Compactification and Extra Dimensions

String theory required extra spatial dimensions—ten in total—for mathematical consistency. These additional dimensions are compactified, meaning they are curled up into tiny shapes that are too small to be observed directly. The process of compactification led to the study of complex geometries known as Calabi-Yau manifolds. These shapes provided the necessary structure for the extra dimensions, influencing the physical properties of the resulting four-dimensional universe.

Dualities

During the first string revolution, physicists discovered various dualities—mathematical relationships that revealed connections between seemingly different string theories. These dualities showed that the five superstring theories were not entirely distinct but were different aspects of the same underlying theory. T-duality, for example, demonstrated that a string propagating in a large circular dimension is equivalent to a string in a small circular dimension, highlighting the deep interconnections within string theory.

Impact and Legacy

The first string revolution established string theory as a major area of research in theoretical physics. It attracted a significant number of physicists to the field, leading to a surge of new ideas and developments. The discovery of anomaly cancellation and the unification of different string theories through dualities provided a solid foundation for future research.

The first string revolution marked a turning point in the quest for a unified theory of fundamental forces. By resolving key mathematical inconsistencies and revealing deep connections between different theories, it paved the way for further advancements in string theory and its quest to describe the underlying fabric of the universe.

Why a Unification Theory in Physics is So Difficult to Achieve

Achieving a unification theory in physics, often referred to as a "Theory of Everything" (ToE), is an extraordinarily challenging goal. This theory aims to unify all fundamental forces and particles into a single comprehensive framework. Here are the main reasons why this goal is so difficult to achieve:

1. Fundamental Forces Disparity

- **Different Natures:** The four fundamental forces—gravity, electromagnetism, the strong nuclear force, and the weak nuclear force—operate on vastly different scales and exhibit different behaviors. Gravity is a geometric property of spacetime, described by general relativity, while the other three forces are described by quantum mechanics through the Standard Model.
- **Mathematical Incompatibility:** General relativity, which describes gravity, is formulated in the language of continuous spacetime, whereas quantum mechanics relies on probabilistic fields and particles. Combining

these frameworks into a single, consistent theory has proven exceptionally difficult due to their fundamentally different mathematical structures.

2. Quantum Gravity

- **Planck Scale:** A unification theory must work at the Planck scale, where quantum effects of gravity become significant. At this scale, the effects of quantum gravity are expected to be dominant, but our current understanding and experimental capabilities are far from reaching this scale.
- **Non-renormalizability:** Traditional methods of dealing with infinities in quantum field theory do not work for gravity. General relativity leads to non-renormalizable infinities, making it impossible to calculate meaningful physical quantities without a new theoretical framework.

3. Complexity and Diversity of Particles

- **Variety of Particles:** The Standard Model of particle physics describes a vast array of particles, including quarks, leptons, and gauge bosons, each with unique properties. Any unification theory must account for all these particles and their interactions, adding to the complexity.
- **Beyond the Standard Model:** The discovery of phenomena that cannot be explained by the Standard Model, such as dark matter, dark energy, and neutrino oscillations, suggests that our current models are incomplete. A ToE must incorporate these unknowns, further complicating its development.

4. Extra Dimensions

- **String Theory:** One of the leading candidates for a ToE, string theory, requires additional spatial dimensions beyond the familiar three. These extra dimensions are hypothesized to be compactified or hidden, making direct experimental verification challenging.
- **Complex Geometries:** The shapes and properties of these extra dimensions, often described by complex geometries like Calabi-Yau manifolds, significantly affect the theory's predictions. Understanding and solving these geometries is a formidable mathematical task.

5. Experimental Limitations

- **Energy Scales:** Testing theories at the energy scales required to probe the fundamental nature of particles and forces (e.g., the Planck scale) is currently beyond the reach of our most powerful particle accelerators.
- **Indirect Evidence:** Much of the evidence for theories beyond the Standard Model, including potential unification theories, must come from indirect observations. These might include subtle effects in high-energy physics experiments, astrophysical observations, or cosmological data, which are often difficult to interpret unambiguously.

6. Mathematical Challenges

- **Advanced Mathematics:** Developing a unification theory involves highly advanced mathematical concepts, many of which are still not fully understood. String theory, for example, relies on complex areas of mathematics such as algebraic geometry, topology, and conformal field theory.
- **Theory Complexity:** The sheer complexity of the mathematical structures involved can make it difficult to derive testable predictions or to even ensure that the theory is internally consistent.

7. Theoretical Diversity

- **Multiple Approaches:** Several competing theories aim to unify fundamental forces, including string theory, loop quantum gravity, and others. Each has its strengths and weaknesses, but none has yet emerged as the definitive solution.
- **Incomplete Frameworks:** Many of these theories are still in development and have unresolved issues. For example, string theory has a vast landscape of possible solutions, making it difficult to pinpoint the correct one that describes our universe.

8. Philosophical and Conceptual Issues

- **Nature of Reality:** Unification theories often lead to deep philosophical questions about the nature of reality, spacetime, and the universe. These questions challenge our fundamental understanding and require not only scientific but also philosophical consideration.
- **Interpretations of Quantum Mechanics:** Different interpretations of quantum mechanics (e.g., Copenhagen interpretation, many-worlds interpretation) offer different views on reality, influencing how a unification theory might be formulated and understood.

Overall, the quest for a unification theory in physics is fraught with profound challenges that span experimental, theoretical, mathematical, and philosophical domains. The disparity between the forces, the complexity of the required mathematics, the limitations of current experimental techniques, and the conceptual shifts needed all contribute to making this one of the most ambitious and difficult goals in science.

CHAPTER 3: FUNDAMENTAL CONCEPTS OF STRING THEORY

Strings and Branes

As we've covered, in string theory, the fundamental building blocks of the universe are not point-like particles but tiny, one-dimensional objects called strings. These strings can vibrate at different frequencies, and each vibration corresponds to a different particle. This idea fundamentally changes our understanding of matter and forces.

Strings
Basic Concept:

- **One-Dimensional Objects:** Strings are minuscule, one-dimensional filaments. Unlike particles, which are zero-dimensional points, strings have length but no other dimensions.
- **Vibrational Modes:** The different vibrational states of strings determine the properties of the particles they represent. For example, one vibration might correspond to an electron, while another might correspond to a photon.
- **Types of Strings:** Strings can be open or closed. Open strings have two distinct endpoints, while closed strings form loops. The type of string affects how it interacts with other strings and branes.

Mathematical Foundation:

- **Action Principle:** The behavior of strings is described by an action principle, similar to how the motion of particles is described in classical mechanics. The Polyakov action is often used to describe string dynamics.
- **Extra Dimensions:** For the mathematics to work, strings require more than the familiar four dimensions of spacetime. The simplest versions of string theory require ten dimensions, while more comprehensive theories like M-theory require eleven dimensions.

Branes
Basic Concept:

- **Multi-Dimensional Objects:** Branes (short for membranes) are multi-dimensional objects that generalize the concept of strings. A p-brane is an object with p spatial dimensions. For example, a 0-brane is a point, a 1-brane is a string, and a 2-brane is a membrane.

- **Types of Branes:** Branes can have various dimensions and can exist within the higher-dimensional space required by string theory. They are important in the dynamics of strings and in the structure of the universe.

Interactions with Strings:

- **Endpoints on Branes:** Open strings can have their endpoints attached to branes. This attachment influences the vibrational modes and properties of the strings.
- **Brane Dynamics:** Branes can move and interact with each other. These interactions can create and annihilate strings, leading to rich and complex physical phenomena.

The Role of Strings and Branes in the Universe

Fundamental Forces:

- **Unification of Forces:** String theory aims to unify all fundamental forces by describing them as different vibrational modes of strings. For example, the graviton, which mediates the force of gravity, is a specific vibrational state of a closed string.
- **Gauge Theories:** The properties of strings and their interactions with branes provide a natural way to describe gauge theories, which are used to explain electromagnetism, the weak force, and the strong force.

Cosmology:

- **Brane World Scenarios:** Some theories suggest that our universe is a 3-brane embedded in a higher-dimensional space. This brane world scenario can provide new insights into cosmology and the nature of the universe.
- **Extra Dimensions:** The compactification of extra dimensions on branes influences the physical laws observed in our four-dimensional spacetime. These compactified dimensions are often modeled using complex geometric shapes like Calabi-Yau manifolds.

Quantum Gravity:

- **Quantum Gravity:** String theory provides a framework for understanding quantum gravity. The quantization of strings naturally incorporates gravitational interactions, offering a potential solution to the problem of unifying quantum mechanics with general relativity.
- **Black Hole Physics:** Branes play a significant role in the study of black holes. The holographic principle, which suggests that the information

within a volume of space can be described by a theory on its boundary, has deep connections to string theory and branes.

Research and Development
Mathematical Techniques:

- **Advanced Mathematics:** The study of strings and branes involves advanced mathematical techniques from areas like algebraic geometry, topology, and conformal field theory. These techniques help physicists understand the properties and interactions of strings and branes.
- **Computational Models:** Researchers use sophisticated computational models to simulate string and brane dynamics. These models help to explore the implications of string theory and to make predictions that can be tested experimentally.

Experimental Challenges:

- **Detection:** Directly detecting strings or branes is beyond current experimental capabilities due to their incredibly small size. Researchers look for indirect evidence, such as the effects of extra dimensions on particle physics and cosmology.
- **Future Experiments:** Advances in technology and experimental techniques may eventually provide ways to test the predictions of string theory more directly, bringing us closer to understanding the true nature of the universe.

Open and Closed Strings

In string theory, the fundamental objects are strings, and they come in two primary forms: open strings and closed strings. Each type has distinct characteristics and has importance in the theory's ability to describe various fundamental particles and forces. Here's an in-depth look at open and closed strings.

Open Strings
Basic Structure:

- **Endpoints:** Open strings have two distinct endpoints. These endpoints can move freely or be attached to objects called branes (short for membranes).
- **Vibrational Modes:** The different vibrational states of an open string determine the properties of the particles they represent. For example, certain vibrational modes of open strings can correspond to gauge bosons, which mediate forces like electromagnetism and the weak nuclear force.

Interactions:

- **Brane Attachment:** Open strings can have their endpoints attached to branes. This attachment can influence the type of particles and interactions that the string represents. For instance, in some string theories, the ends of open strings are confined to specific branes, leading to a rich structure of possible interactions.
- **Forces and Fields:** Open strings are often associated with gauge fields, which describe the fundamental forces in particle physics. When the endpoints of open strings move on a brane, they can generate gauge fields, leading to interactions that are consistent with the Standard Model of particle physics.

Mathematical Description:

- **Worldsheet:** The two-dimensional surface traced out by an open string as it moves through spacetime is called a worldsheet. The endpoints of the open string trace out boundary lines on this worldsheet.
- **Boundary Conditions:** The behavior of open strings is influenced by boundary conditions at their endpoints. These conditions are essential for determining the allowed vibrational modes and, consequently, the physical properties of the strings.

Closed Strings

Basic Structure:

- **Loops:** Closed strings are loops with no endpoints. They can move freely through the higher-dimensional space proposed by string theory.
- **Vibrational Modes:** The vibrational modes of closed strings correspond to different particles, including those that mediate gravity, such as gravitons. The closed loop structure allows for more complex vibrational patterns compared to open strings.

Interactions:

- **Self-Interaction:** Closed strings can interact with themselves by splitting into two strings or joining to form a single string. These interactions are fundamental to the dynamics of closed strings and lead to the rich behavior seen in string theory.
- **Gravitational Force:** One of the key vibrational modes of a closed string corresponds to the graviton, the hypothetical particle that mediates the force of gravity. This makes closed strings central to the description of gravitational interactions in string theory.

Mathematical Description:

- **Worldsheet:** The worldsheet of a closed string is a cylindrical surface as it moves through spacetime. This closed surface structure is crucial for understanding the interactions and properties of closed strings.
- **No Boundary Conditions:** Unlike open strings, closed strings do not have endpoints, so they do not require boundary conditions. This difference simplifies some aspects of their mathematical description but also introduces new complexities in their interactions.

Combined Dynamics

Interaction between Open and Closed Strings:

- **String Interactions:** Open and closed strings can interact in various ways. For example, an open string can form a closed string by joining its endpoints, or a closed string can break into an open string with its endpoints attached to branes.
- **Unified Framework:** These interactions provide a unified framework for understanding different types of particles and forces. For instance, the gauge fields associated with open strings and the gravitational fields associated with closed strings can interact within this framework, offering a potential path to unifying all fundamental forces.

Physical Implications

Particle Physics:

- **Gauge Bosons:** Open strings with endpoints on branes can describe gauge bosons, which are particles that mediate the fundamental forces such as electromagnetism, the weak force, and the strong force.
- **Gravitons:** Closed strings naturally incorporate gravitons, which are necessary for a quantum theory of gravity. This inclusion makes string theory a strong candidate for a unified theory of all fundamental forces.

Cosmology and Extra Dimensions:

- **Brane Worlds:** The idea that open strings have their endpoints attached to branes leads to the concept of brane worlds, where our observable universe could be a three-dimensional brane embedded in a higher-dimensional space.
- **Compactified Dimensions:** Both open and closed strings propagate through the extra dimensions required by string theory. The compactification of these dimensions, often into shapes like Calabi-Yau manifolds, influences the physical properties of the strings and the resulting particles.

Research and Challenges

Experimental Detection:

- **Energy Scales:** The typical energy scales required to directly detect strings are far beyond current experimental capabilities. Indirect evidence, such as effects on high-energy particle collisions or cosmological observations, is being sought to support string theory.
- **Theoretical Models:** Researchers continue to develop theoretical models to better understand the implications of string theory and to make predictions that can be tested with existing and future technology.

Vibrations and Modes

In string theory, the concept of vibrations and modes is central to understanding how strings give rise to different particles. Here's a detailed look at these concepts:

Vibrations of Strings

Basic Idea:

- **One-Dimensional Objects:** Strings are tiny, one-dimensional objects that can vibrate in various ways. The specific pattern of a string's vibration determines the type of particle it represents.
- **Frequency and Energy:** Just like a musical string can produce different notes depending on its frequency, a string in string theory vibrates at different frequencies, with each frequency corresponding to a different energy level.

Vibrational Modes

Definition:

- **Modes:** The specific patterns or ways in which a string vibrates are called modes. Each mode corresponds to a different state of the string, which can be associated with a different particle.
- **Harmonics:** Similar to musical instruments, strings in string theory can vibrate in fundamental modes and harmonics. The fundamental mode is the simplest vibration, while harmonics are more complex patterns of vibration.

Types of Vibrations:

- **Longitudinal Modes:** Vibrations along the length of the string. These are the simplest forms of vibrations.

- **Transverse Modes:** Vibrations perpendicular to the length of the string. These can occur in multiple directions, especially in higher-dimensional spaces required by string theory.
- **Rotational and Oscillatory Modes:** More complex vibrations involving rotations and oscillations in multiple dimensions.

Mathematical Description

Action Principle:

- **Polyakov Action:** The dynamics of string vibrations are often described using the Polyakov action, which generalizes the action principle used in classical mechanics and quantum field theory. This action describes how strings propagate through spacetime and how their vibrations evolve.
- **Equations of Motion:** The equations of motion derived from the action principle govern how strings vibrate. These equations are solved to find the allowed vibrational modes and their corresponding energies.

Particle Correspondence

Mass and Spin:

- **Massless and Massive Particles:** The different vibrational modes of strings correspond to particles with varying masses. Some modes represent massless particles like photons, while others correspond to massive particles.
- **Spin:** The spin of a particle, a fundamental property related to its angular momentum, is also determined by the string's vibrational mode. Different modes can represent particles with different spins, such as fermions (with half-integer spins) and bosons (with integer spins).

Graviton and Other Particles:

- **Graviton:** In closed string theory, one of the vibrational modes of a closed string corresponds to the graviton, the hypothetical quantum particle that mediates gravity. This is a crucial aspect of string theory, as it naturally includes gravity.
- **Gauge Bosons:** Open strings can have modes corresponding to gauge bosons, which mediate forces like electromagnetism and the weak and strong nuclear forces.

Higher Dimensions and Compactification

Extra Dimensions:

- **Requirement for Consistency:** For string theory to be mathematically consistent, it requires more than the familiar four dimensions of spacetime. Typically, ten or eleven dimensions are needed.
- **Compactification:** The extra dimensions are compactified, meaning they are curled up into tiny, unobservable shapes. The geometry of these compactified dimensions affects the vibrational modes of strings.

Calabi-Yau Manifolds:

- **Shape and Influence:** Calabi-Yau manifolds are complex geometric shapes used to model the compactified dimensions in string theory. The specific shape of these manifolds influences the allowed vibrational modes and, therefore, the properties of particles.

Quantum Mechanics and String Vibrations

Quantization:

- **Quantum States:** The vibrational modes of strings are quantized, meaning only specific discrete energy levels are allowed. This is analogous to the quantization of energy levels in an atom.
- **Probability Amplitudes:** The behavior of strings is described using probability amplitudes, similar to other quantum systems. These amplitudes determine the likelihood of a string being found in a particular vibrational mode.

Superposition:

- **Multiple Modes:** A string can exist in a superposition of multiple vibrational modes, just as particles can exist in a superposition of quantum states. This leads to the rich and complex behavior seen in string theory.

Implications and Research

Unification of Forces:

- **Unified Framework:** The different vibrational modes of strings provide a unified framework for describing all fundamental particles and forces. This unification is one of the key motivations behind string theory.
- **Potential Discoveries:** Ongoing research aims to understand how different modes correspond to known particles and to predict new particles and interactions that could be experimentally verified.

Challenges:

- **Mathematical Complexity:** The mathematics involved in describing string vibrations and their modes is highly complex, requiring advanced tools from algebraic geometry, topology, and quantum field theory.
- **Experimental Evidence:** Directly observing string vibrations is beyond current experimental capabilities due to their incredibly small scale. Researchers seek indirect evidence through high-energy physics experiments and cosmological observations.

String Interactions

Interactions between strings are fundamental processes that explain how particles interact with each other. Unlike point particles, which interact at specific points in space and time, strings interact by joining and splitting in ways that spread out over a small but finite area. This difference in interaction mechanisms helps address some of the mathematical issues that arise in particle physics. Here's a detailed look at string interactions:

Basic Concepts of String Interactions

1. Joining and Splitting:

- **Joining:** Two strings can join together to form a single string. This process can happen for both open and closed strings. When two open strings join, their endpoints meet and merge. When two closed strings join, they form a single loop.
- **Splitting:** A single string can split into two strings. This can occur for open strings, where the string breaks at a point, creating two new endpoints, or for closed strings, where the loop splits into two separate loops.

2. Interaction Diagrams:

- **Worldsheet:** The interactions of strings are described by a two-dimensional surface called a worldsheet. This surface traces the history of the string's motion through spacetime. When strings interact, the worldsheet evolves accordingly, showing the process of joining or splitting.
- **Feynman Diagrams:** In traditional quantum field theory, particle interactions are depicted using Feynman diagrams. String interactions can also be represented by similar diagrams, but they involve surfaces rather than points. These diagrams help visualize how strings interact over time.

Types of Interactions

1. Open String Interactions:

- **Endpoint Attachment:** Open strings often have their endpoints attached to branes. The interactions between open strings can lead to the creation or annihilation of strings on these branes, affecting the fields and forces that the strings represent.
- **Gauge Interactions:** The endpoints of open strings can carry charges, and their interactions can correspond to the exchange of gauge bosons, which mediate forces like electromagnetism and the weak and strong nuclear forces.

2. Closed String Interactions:

- **Gravitational Interactions:** One of the key vibrational modes of a closed string corresponds to the graviton, the hypothetical particle that mediates gravity. When closed strings interact, they can exchange gravitons, leading to gravitational interactions.
- **Self-Interaction:** Closed strings can interact with themselves by splitting into two strings or joining from two into one. These self-interactions are fundamental in understanding the dynamics of closed strings and the forces they mediate.

Mathematical Framework

1. String Field Theory:

- **Action and Equations of Motion:** String field theory provides a framework for describing string interactions using fields, similar to how quantum field theory describes particle interactions. The action principle and resulting equations of motion govern how strings evolve and interact.
- **Path Integrals:** The path integral formulation, used in quantum mechanics and quantum field theory, is extended to string theory. This approach sums over all possible histories of the strings, including all possible ways they can join and split, to calculate interaction probabilities.

2. Conformal Field Theory:

- **Worldsheet Description:** The behavior of strings on the worldsheet is described using conformal field theory. This mathematical framework helps understand how strings interact and how their vibrational modes correspond to different particles.
- **Vertex Operators:** In conformal field theory, vertex operators are used to insert interactions into the worldsheet. These operators help describe how strings interact at specific points on the worldsheet.

Phenomenological Implications

1. Particle Interactions:

- **Force Mediation:** String interactions naturally incorporate the mediation of fundamental forces. For example, the exchange of vibrational modes corresponding to gauge bosons or gravitons explains how forces like electromagnetism and gravity operate.
- **Unification:** The framework of string interactions provides a unified description of all fundamental forces, potentially leading to a Theory of Everything (ToE). This unification is one of the primary motivations behind string theory.

2. Supersymmetry:

- **Superpartners:** String interactions often involve supersymmetry, where each particle has a corresponding superpartner with different spin properties. Supersymmetry helps stabilize the theory and resolve certain theoretical problems, such as the hierarchy problem.
- **Extended Interactions:** Supersymmetric string theories include interactions between superpartners, leading to richer and more complex dynamics.

Challenges and Future Directions

1. Mathematical Complexity:

- **Advanced Techniques:** The mathematics of string interactions is highly complex, involving advanced techniques from algebraic geometry, topology, and conformal field theory. Researchers continue to develop new methods to better understand and solve these complex equations.
- **Computational Models:** Numerical simulations and computational models be important in exploring string interactions. These models help visualize and test theoretical predictions, providing insights into the dynamics of strings.

2. Experimental Verification:

- **Indirect Evidence:** Directly observing string interactions is beyond current experimental capabilities due to the incredibly small scale of strings. However, researchers look for indirect evidence, such as effects on high-energy particle collisions or cosmological observations, to support string theory.
- **Future Experiments:** Advances in technology and experimental techniques may eventually provide ways to test the predictions of string theory more directly. High-energy experiments and precision measurements in cosmology could offer clues about the validity of string interactions.

CHAPTER 4: THE MATHEMATICAL FOUNDATIONS

Basic Mathematical Concepts & Techniques

String theory relies on a sophisticated mathematical framework to describe the fundamental nature of the universe. Here, we'll explore some of the basic mathematical concepts and techniques essential for understanding string theory.

1. Strings and Worldsheet

Strings as Fundamental Objects:

- **One-Dimensional Strings:** In string theory, particles are modeled as tiny, one-dimensional strings. These strings can vibrate in different modes, with each mode representing a different particle.
- **Worldsheet:** The path a string traces out in spacetime is called its worldsheet. For an open string, this worldsheet looks like a strip, while for a closed string, it resembles a tube or cylinder.

Action Principle:

- **Polyakov Action:** The behavior of strings is described by the Polyakov action, an extension of the action principle used in classical mechanics. It helps determine how the worldsheet evolves through spacetime.

2. Extra Dimensions

Higher-Dimensional Space:

- **Requirement for Consistency:** String theory requires more than the familiar four dimensions (three of space and one of time). The simplest versions of string theory need ten dimensions for mathematical consistency.
- **Compactification:** The extra dimensions are compactified, meaning they are curled up into very small shapes that are difficult to detect. These compactified dimensions influence the physical properties of the strings.

Calabi-Yau Manifolds:

- **Complex Geometry:** The shapes of these extra dimensions are often modeled using Calabi-Yau manifolds, which are complex, multi-dimensional geometric structures. These manifolds ensure the theory's consistency and preserve certain symmetries.

3. Conformal Field Theory

Worldsheet Dynamics:

- **Field Theory on the Worldsheet:** The physics of the string's worldsheet is described by conformal field theory (CFT). CFT studies fields that are invariant under conformal transformations, which preserve angles but not necessarily distances.
- **Vertex Operators:** These operators are used within CFT to insert interactions on the worldsheet. They help describe how strings interact at specific points.

4. Supersymmetry

Symmetry Between Particles:

- **Supersymmetry (SUSY):** This theoretical symmetry pairs each particle with a superpartner that has a different spin. SUSY helps solve several theoretical issues, such as the hierarchy problem, by canceling out certain mathematical infinities.
- **Superstring Theory:** Incorporating SUSY into string theory leads to superstring theory, which requires ten dimensions and provides a more stable theoretical framework.

5. Dualities

Mathematical Equivalences:

- **S-Duality and T-Duality:** These are types of dualities that reveal deep connections between seemingly different string theories. S-duality relates strong and weak coupling constants, while T-duality connects large and small distance scales.
- **Unified Framework:** Dualities show that the five different superstring theories are actually different aspects of a single underlying theory, often referred to as M-theory in the context of eleven dimensions.

6. Path Integrals

Quantum Mechanics and Strings:

- **Summing Over Histories:** The path integral formulation of quantum mechanics is extended to string theory. It involves summing over all possible histories of the strings, including all the ways they can join and split.
- **Amplitude Calculations:** Path integrals are used to calculate the probability amplitudes for various string interactions, which determine the likelihood of different physical processes.

7. Moduli Space

Parameter Space of Solutions:

- **Defining Moduli Space:** This space encompasses all possible shapes and sizes of the compactified dimensions. Each point in moduli space represents a different way to compactify the extra dimensions.
- **Physical Implications:** The properties of particles and forces depend on the location within moduli space, as different compactifications lead to different physical laws.

8. Algebraic Geometry

Describing Extra Dimensions:

- **Complex Shapes:** Algebraic geometry provides the tools to describe the complex shapes of the compactified dimensions. Techniques from this field are essential for solving the equations that arise in string theory.
- **Topological Invariants:** These invariants help classify the different possible shapes of the extra dimensions and determine their physical properties.

9. Topology

Properties of Space:

- **Topological Structures:** Topology studies properties of space that remain unchanged under continuous deformations. In string theory, topology helps understand the global properties of the extra dimensions.
- **String Interactions:** Topological considerations are crucial in understanding how strings can interact, join, and split within the higher-dimensional space.

10. Holographic Principle

Connecting Dimensions:

- **AdS/CFT Correspondence:** Proposed by Juan Maldacena, this principle suggests a relationship between a string theory in a higher-dimensional anti-de Sitter (AdS) space and a conformal field theory on its lower-dimensional boundary. This duality provides insights into quantum gravity.

Complex Numbers and Their Role

Complex numbers are fundamental to string theory, providing a mathematical foundation for many of its core concepts and techniques. At their most basic, complex numbers are numbers of the form z = a + bi, where a and b are real numbers, and i is the imaginary unit with the property that $i^2 = -1$.

In string theory, complex numbers have several important roles:

1. **Describing the String Worldsheet**: The two-dimensional surface traced by a string (called the worldsheet) is often described using complex coordinates. This simplifies the mathematical treatment of the worldsheet's geometry and dynamics.
2. **Conformal Field Theory (CFT)**: In CFT, which is essential to string theory, fields on the worldsheet can be decomposed into holomorphic and anti-holomorphic components using complex coordinates.
3. **String Interactions**: Complex numbers are used in calculating string interaction amplitudes. The path integral formulation of string theory often involves integrating over complex planes.
4. **Extra Dimensions**: The extra dimensions in string theory are often compactified into shapes called Calabi-Yau manifolds, which are complex manifolds. Complex coordinates are essential for describing their properties.
5. **Supersymmetry**: Complex numbers are used to express superfields, which incorporate both bosonic and fermionic components in supersymmetric theories.
6. **Dualities**: Various dualities in string theory, such as S-duality and T-duality, often involve complex mappings that transform one theory into another.
7. **Quantum Mechanics**: The wavefunctions describing quantum states of strings are complex-valued, with probability amplitudes given by the square of their modulus.

Advanced techniques in string theory, such as contour integration and the residue theorem from complex analysis, further emphasize the importance of complex numbers in this field.

In essence, complex numbers provide a mathematical toolkit that allows string theorists to describe and manipulate the intricate structures and dynamics of strings in multiple dimensions. Their use simplifies many calculations and reveals deep connections within the theory.

Symmetries in String Theory

Symmetries have a fundamental role in string theory, shaping the framework and guiding the interactions and behaviors of strings. Understanding these symmetries is crucial for grasping the deeper structure and implications of the theory. Here's an in-depth look at the key symmetries in string theory.

1. Poincaré Symmetry

Space-Time Symmetry:

- **Poincaré Group:** The Poincaré group is the symmetry group of Minkowski spacetime, encompassing translations, rotations, and boosts. It describes the invariance of the laws of physics under these transformations.
- **String Dynamics:** The equations governing string dynamics are invariant under Poincaré transformations, ensuring that the theory respects the basic symmetries of spacetime.

2. Conformal Symmetry

Worldsheet Symmetry:

- **Conformal Group:** Conformal symmetry refers to the invariance of the worldsheet theory under transformations that preserve angles but not necessarily distances. This symmetry is described by the conformal group.
- **Virasoro Algebra:** The local conformal symmetry on the worldsheet is captured by the Virasoro algebra, which is an infinite-dimensional algebra crucial for the consistency of string theory.
- **CFT:** Conformal field theory (CFT) describes the dynamics of the string's worldsheet. The use of complex coordinates simplifies the mathematical treatment, leveraging the rich structure of conformal symmetry.

3. Supersymmetry

Symmetry Between Fermions and Bosons:

- **SUSY:** Supersymmetry is a theoretical symmetry that relates bosons (particles that mediate forces) and fermions (particles that make up matter). Each particle has a corresponding superpartner with a different spin.
- **Superstring Theory:** Incorporating supersymmetry into string theory leads to superstring theory, which requires ten dimensions for mathematical consistency. This framework helps resolve certain theoretical issues and provides a more stable theory.
- **Superfields:** In supersymmetric theories, superfields are used to combine bosonic and fermionic components. These fields transform under supersymmetry operations, maintaining the symmetry of the theory.

4. Gauge Symmetry

Internal Symmetry:

- **Gauge Groups:** Gauge symmetries involve transformations that vary from point to point in spacetime. In string theory, these symmetries are associated with the different types of interactions and forces, described by gauge groups like SU(N), SO(N), and E8.

- **Open Strings:** The endpoints of open strings can carry charges and interact with gauge fields. The gauge symmetry determines the allowed interactions and the structure of the gauge fields.

5. T-Duality and S-Duality

Duality Symmetries:

- **T-Duality:** T-duality is a symmetry that relates string theories compactified on circles of radius RRR to theories compactified on circles of radius $1/R1/R1/R$. This duality shows that strings on large and small scales can be equivalent, revealing deep connections between seemingly different theories.
- **S-Duality:** S-duality relates strong coupling regimes of a theory to the weak coupling regimes of another. It indicates that the behavior of strings at high interaction strengths can be mapped to the behavior at low interaction strengths in a different framework.

6. Global Symmetries

Discrete and Continuous Symmetries:

- **Discrete Symmetries:** These include symmetries like charge conjugation (C), parity (P), and time reversal (T). The combined CPT symmetry is a fundamental symmetry in quantum field theory and string theory.
- **Continuous Symmetries:** Apart from gauge symmetries, string theory can exhibit continuous global symmetries, which involve transformations that do not depend on spacetime points.

7. Modular Invariance

String Compactification:

- **Modular Group:** When strings are compactified on tori or other complex shapes, the resulting theory must be invariant under the modular group, which includes transformations that mix and rescale the compactification parameters.
- **Consistency Condition:** Modular invariance is crucial for the consistency of the string theory on these compactified spaces, ensuring that physical predictions do not depend on arbitrary choices of parameters.

8. Enhanced Symmetries in Higher Dimensions

String Compactification:

- **Kaluza-Klein Theory:** Compactification of extra dimensions can lead to enhanced symmetries in the lower-dimensional effective theory. For instance, compactifying on specific manifolds can lead to additional gauge symmetries or global symmetries.
- **Calabi-Yau Manifolds:** Compactifying string theory on Calabi-Yau manifolds can result in supersymmetric field theories with rich symmetry structures that are crucial for describing the physical properties of particles and interactions in four dimensions.

9. Brane Symmetries

Extended Objects:

- **D-Branes:** D-branes are extended objects in string theory that can have various dimensions. The presence of D-branes breaks some of the symmetries of the bulk theory but introduces new symmetries related to the branes themselves.
- **Gauge Symmetries on Branes:** The dynamics of open strings attached to D-branes give rise to gauge theories on the branes' worldvolume, leading to additional gauge symmetries that play a vital role in the theory.

Mathematical Techniques Leveraging Symmetry

Algebraic Structures:

- **Lie Algebras:** Many symmetries in string theory are described using Lie algebras, which provide a mathematical framework for continuous symmetry groups. The representations of these algebras help classify particles and interactions.
- **Group Theory:** Group theory is essential for understanding the structure of symmetries in string theory, including gauge groups, dualities, and discrete symmetries.

Differential Geometry and Topology:

- **Manifold Structures:** The geometric and topological properties of the extra dimensions, often described using complex manifolds like Calabi-Yau spaces, are crucial for understanding how symmetries manifest in the compactified theory.
- **Bundle Theory:** Gauge fields and their symmetries are often described using fiber bundles, a concept from differential geometry that helps model how fields transform under symmetry operations.

Symmetries in string theory are essential for its mathematical consistency and physical implications. They provide a unified framework for describing the interactions and properties of strings, guiding the development of the theory and its connection to observable phenomena. Understanding these symmetries offers profound insights into the fundamental structure of the universe.

Calabi-Yau Manifolds

Calabi-Yau manifolds are important in string theory, particularly in the context of compactification, where extra dimensions are compactified into complex shapes to reconcile the theory with our four-dimensional observable universe. Here's an in-depth look at Calabi-Yau manifolds and their significance in string theory.

What Are Calabi-Yau Manifolds?

Definition:

- **Complex Manifolds:** Calabi-Yau manifolds are complex, multi-dimensional shapes with specific properties that make them suitable for compactifying extra dimensions in string theory. They are named after mathematicians Eugenio Calabi and Shing-Tung Yau.
- **Ricci-Flat:** One of the defining properties of Calabi-Yau manifolds is that they are Ricci-flat, meaning they have zero Ricci curvature. This property ensures that they can support supersymmetry, a critical feature in string theory.

Dimensions:

- **Six-Dimensional Spaces:** In the context of string theory, Calabi-Yau manifolds are typically six-dimensional. When combined with the four familiar dimensions of spacetime, this makes up the ten dimensions required by superstring theory.

Role in String Theory

Compactification:

- **Extra Dimensions:** String theory posits that the universe has more than the familiar four dimensions (three spatial and one temporal). The additional six dimensions are compactified into a Calabi-Yau manifold, making them small and unobservable at everyday scales.
- **Shape and Size:** The specific shape and size of the Calabi-Yau manifold influence the physical properties of the universe. These shapes determine the possible vibrational modes of strings, which correspond to different particles and forces.

Preserving Supersymmetry:

- **Ricci-Flat Geometry:** The Ricci-flat property of Calabi-Yau manifolds allows them to preserve a portion of the supersymmetry in the four-dimensional effective theory. This is crucial for maintaining the stability and consistency of the theory.
- **Holonomy Group:** The holonomy group of a Calabi-Yau manifold is SU(3), which helps in preserving supersymmetry when dimensions are compactified. This group is a key aspect of the manifold's geometric structure.

Mathematical Properties

Complex Geometry:

- **Kähler Manifolds:** Calabi-Yau manifolds are a subset of Kähler manifolds, which have a complex structure compatible with a Riemannian metric. This means they have a rich geometric and algebraic structure that facilitates mathematical analysis.
- **Holomorphic Forms:** These manifolds possess a holomorphic volume form, which remains invariant under the manifold's complex structure. This form is essential for defining the manifold's properties and for performing calculations in string theory.

Topological Features:

- **Betti Numbers:** Betti numbers are topological invariants that describe the number of independent cycles of different dimensions within a manifold. In Calabi-Yau manifolds, these numbers are important in determining the manifold's properties.
- **Euler Characteristic:** The Euler characteristic is another topological invariant that provides information about the manifold's shape and structure. For Calabi-Yau manifolds, the Euler characteristic helps in understanding the compactification and the resulting physical implications.

Physical Implications

Particle Physics:

- **Massless Modes:** The specific geometry of a Calabi-Yau manifold determines the massless modes of the compactified theory. These modes correspond to the particles observed in the four-dimensional universe.
- **Gauge Symmetry:** The structure of the manifold can lead to various gauge symmetries in the effective four-dimensional theory. This is crucial for describing the fundamental forces within the framework of string theory.

String Vacua:

- **Landscape of Vacua:** The vast number of possible Calabi-Yau manifolds contributes to the string theory landscape, a multitude of possible vacuum states. Each vacuum corresponds to a different configuration of the compactified dimensions and can lead to different physical laws.
- **Moduli Space:** The parameters describing the shape and size of the Calabi-Yau manifold form a moduli space. The dynamics within this space determine the properties of the resulting four-dimensional universe, including the values of physical constants.

Examples and Applications

Specific Manifolds:

- **Quintic Threefold:** One of the simplest examples of a Calabi-Yau manifold is the quintic threefold, defined by a polynomial equation in complex projective space. This manifold is often used in explicit calculations and model building within string theory.
- **Toric Varieties:** Calabi-Yau manifolds can also be constructed using toric geometry, which provides a combinatorial approach to their classification and study.

Mirror Symmetry:

- **Dual Manifolds:** Mirror symmetry is a phenomenon where pairs of Calabi-Yau manifolds, known as mirror pairs, lead to equivalent physical theories. This symmetry provides powerful tools for studying these manifolds and their implications in string theory.
- **Mathematical Insights:** Mirror symmetry has led to significant advancements in both mathematics and physics, providing deep insights into the geometry of Calabi-Yau manifolds and their role in string compactification.

Research and Challenges

Mathematical Complexity:

- **Advanced Techniques:** The study of Calabi-Yau manifolds requires advanced mathematical techniques from algebraic geometry, differential geometry, and topology. Researchers continuously develop new methods to explore their properties and implications.
- **Computational Models:** Numerical simulations and computational models play a vital role in understanding the geometry and physics of Calabi-Yau manifolds. These models help visualize the complex shapes and analyze their effects on string theory.

Experimental Verification:

- **Indirect Evidence:** Direct experimental evidence for the specific shapes of Calabi-Yau manifolds is currently beyond reach. However, researchers look for indirect signs, such as specific patterns in particle physics experiments and cosmological observations.
- **Future Prospects:** Advances in experimental techniques and technology may provide new ways to test the predictions of string theory and the role of Calabi-Yau manifolds, potentially offering insights into the fundamental nature of the universe.

Calabi-Yau manifolds are essential to string theory, providing a means to compactify extra dimensions and determine the physical properties of the resulting four-dimensional universe. Their complex geometric and topological structures are crucial for preserving supersymmetry and enabling a unified description of fundamental particles and forces. Understanding these manifolds is a key challenge and a central focus in the ongoing development of string theory.

CHAPTER 5: STRING THEORY AND QUANTUM MECHANICS

Unifying Principles

String theory and quantum mechanics are deeply interconnected, with string theory providing a framework that extends and unifies the principles of quantum mechanics to describe the fundamental nature of the universe. Here's an exploration of the unifying principles of string theory and quantum mechanics.

Quantum Mechanics: The Foundation
Wave-Particle Duality:

- **Particles as Waves:** In quantum mechanics, particles exhibit both wave-like and particle-like properties. This duality is fundamental to understanding phenomena at microscopic scales.
- **Probability Waves:** The state of a particle is described by a wavefunction, which provides the probabilities of finding the particle in different positions and states.

Quantization:

- **Discrete Energy Levels:** Quantum mechanics introduces the concept of quantization, where certain properties, such as energy, can only take on discrete values. This is evident in the energy levels of electrons in an atom.
- **Operators and Commutators:** Physical observables are represented by operators. The commutation relations between these operators encode the fundamental uncertainties and constraints of the system.

Uncertainty Principle:

- **Heisenberg's Principle:** The Heisenberg Uncertainty Principle states that it is impossible to simultaneously know both the exact position and momentum of a particle. This principle introduces inherent uncertainties into the behavior of quantum systems.

Extending to String Theory
Strings as Fundamental Entities:

- **Vibrating Strings:** String theory replaces point-like particles with one-dimensional strings. These strings can vibrate at different frequencies, with each vibrational mode corresponding to a different particle.

- **Multiple Dimensions:** Strings require additional spatial dimensions for consistency. In the simplest versions of string theory, there are ten dimensions, while M-theory suggests eleven dimensions.

Unified Framework:

- **Inclusion of Gravity:** Unlike quantum field theory, which successfully describes the other three fundamental forces (electromagnetism, weak, and strong nuclear forces), string theory naturally includes gravity. The graviton, the hypothetical quantum particle that mediates gravity, arises as a vibrational mode of a closed string.
- **Supersymmetry:** String theory often incorporates supersymmetry, a symmetry that pairs each particle with a superpartner differing in spin. Supersymmetry helps cancel out certain mathematical inconsistencies and provides a more coherent framework.

Mathematical Principles

Action Principle:

- **Polyakov Action:** The dynamics of strings are governed by the Polyakov action, an extension of the classical action principle used in quantum mechanics. This action describes how strings propagate through spacetime and how their vibrations evolve.

Conformal Field Theory:

- **Worldsheet Dynamics:** The two-dimensional surface traced out by a string, called the worldsheet, is described using conformal field theory (CFT). CFT helps analyze the string's vibrations and interactions by leveraging the rich symmetry structure of the worldsheet.

Path Integrals:

- **Summing Over Histories:** In both quantum mechanics and string theory, the path integral formulation is used to calculate probabilities by summing over all possible histories of the system. In string theory, this involves integrating over all possible configurations of the string worldsheet.

Unifying Quantum Mechanics and General Relativity

Quantum Gravity:

- **Graviton Emergence:** String theory provides a framework for quantum gravity by naturally incorporating the graviton. This unifies the principles

of quantum mechanics with general relativity, addressing one of the biggest challenges in theoretical physics.
- **Smooth Interactions:** Strings interact over a finite area, which helps avoid the singularities that plague point-particle theories. This smooth interaction structure is crucial for developing a consistent theory of quantum gravity.

Physical Implications

Particle Spectrum:

- **Diverse Particles:** The different vibrational modes of strings correspond to the spectrum of particles observed in nature. This includes not only the familiar particles of the Standard Model but also potentially undiscovered particles predicted by the theory.

Force Unification:

- **Single Framework:** String theory aims to describe all fundamental forces within a single theoretical framework. The interactions between strings naturally produce the gauge bosons that mediate the electromagnetic, weak, and strong forces, alongside gravity.

Mathematical Tools

Advanced Geometry:

- **Calabi-Yau Manifolds:** The extra dimensions in string theory are often compactified into complex shapes called Calabi-Yau manifolds. These manifolds provide the necessary conditions for preserving supersymmetry and determining the physical properties of the compactified theory.
- **Topology and Algebraic Geometry:** Tools from topology and algebraic geometry are essential for understanding the complex structures and interactions in string theory.

Dualities:

- **S-Duality and T-Duality:** These dualities reveal deep connections between different string theories, showing that they are different aspects of a single underlying theory. S-duality relates strong and weak coupling regimes, while T-duality connects large and small distance scales.

Quantum Gravity

Quantum gravity is the field of theoretical physics that seeks to describe gravity according to the principles of quantum mechanics. Traditional theories like general relativity describe gravity as a curvature of spacetime caused by mass and energy, but they do not incorporate quantum mechanics.

The Challenge of Quantum Gravity

Incompatibility of Theories:

- **General Relativity:** Describes gravity on large scales, such as stars, planets, and galaxies. It treats gravity as a continuous field and spacetime as a smooth, curved manifold.
- **Quantum Mechanics:** Describes the other fundamental forces— electromagnetism, the weak nuclear force, and the strong nuclear force— on the smallest scales, where particles exhibit both wave and particle properties. It relies on probabilistic events and discrete energy levels.

Unifying These Frameworks:

- **Singularities and Infinities:** When trying to apply quantum mechanics to general relativity, equations often produce infinities and singularities, such as those found at the centers of black holes or the Big Bang. These infinities suggest that a new framework is needed to describe gravity at quantum scales.

String Theory and Quantum Gravity

Strings as Fundamental Entities:

- **Vibrating Strings:** In string theory, the fundamental objects are tiny, one-dimensional strings. Their different vibrational modes correspond to different particles, including the graviton, the hypothetical quantum particle that mediates gravity.
- **Finite Size:** Strings have a finite length, which helps avoid the problematic infinities that arise when particles are considered as points.

Graviton Emergence:

- **Closed Strings:** The graviton naturally emerges as a vibrational mode of closed strings in string theory. This mode is massless and has a spin of 2, properties that match the theoretical requirements for a particle that mediates the gravitational force.
- **Unified Description:** This inclusion of the graviton within string theory provides a unified description of all fundamental forces, integrating gravity with the quantum framework used for the other three forces.

Mathematical Framework

Action Principle:

- **Polyakov Action:** The behavior of strings is described by the Polyakov action, an extension of the classical action principle used in quantum mechanics. This action governs how strings move and interact in spacetime.
- **Worldsheet:** The two-dimensional surface traced out by a string as it moves through spacetime is called the worldsheet. This concept replaces the point particle's trajectory and is crucial for describing string interactions.

Path Integrals:

- **Summing Over Histories:** In quantum mechanics, path integrals sum over all possible paths a particle can take. In string theory, this concept extends to summing over all possible worldsheet configurations. This approach helps calculate probabilities for different string interactions.

Extra Dimensions and Compactification

Higher Dimensions:

- **Ten Dimensions:** For mathematical consistency, string theory requires ten dimensions. The extra six dimensions are compactified, meaning they are curled up into tiny shapes that are not observable at everyday scales.
- **Calabi-Yau Manifolds:** These compactified dimensions are often modeled using Calabi-Yau manifolds, which have specific properties that preserve supersymmetry and determine the vibrational modes of strings.

Dualities

Symmetry Relations:

- **T-Duality:** Relates theories compactified on large and small radii, showing that strings propagating in a large dimension are equivalent to strings in a small dimension.
- **S-Duality:** Connects strong and weak coupling regimes of string theories, indicating that the behavior of strings at strong interactions can be mapped to their behavior at weak interactions. These dualities reveal that seemingly different string theories are actually different perspectives on the same underlying theory.

Implications for Black Holes and the Early Universe

Black Hole Information Paradox:

- **String Theory Solutions:** String theory provides insights into the black hole information paradox, suggesting that information is not lost in black holes but encoded in the vibrations of strings.
- **Holographic Principle:** The AdS/CFT correspondence, a conjecture in string theory, proposes that a higher-dimensional gravitational theory can be described by a lower-dimensional conformal field theory on the boundary of the space. This principle offers a new way to understand black holes and quantum gravity.

Cosmology:

- **Early Universe:** String theory has implications for the early universe, including the nature of the Big Bang and cosmic inflation. The framework can potentially explain the initial conditions that led to the formation of the universe as we observe it today.

Current Research and Challenges

Experimental Evidence:

- **Indirect Tests:** Direct detection of strings or gravitons is beyond current experimental capabilities. Researchers seek indirect evidence through high-energy physics experiments, cosmological observations, and the study of black holes.
- **Technological Advances:** Future advancements in technology and experimental techniques may provide new ways to test the predictions of string theory and its description of quantum gravity.

Theoretical Developments:

- **Mathematical Rigor:** String theory continues to evolve with new mathematical tools and techniques, enhancing our understanding of its implications for quantum gravity.
- **Interdisciplinary Impact:** Insights from string theory contribute to various fields, including condensed matter physics, quantum computing, and cosmology, highlighting its broad impact on our understanding of the fundamental nature of reality.

Quantum Fields and Strings

To fully appreciate how strings relate to quantum fields, it's essential to understand the concepts of quantum fields and how they are extended in string theory.

Quantum Fields

Definition:

- **Quantum Field Theory (QFT):** Quantum fields are the fundamental entities in quantum field theory. Each type of particle is associated with a corresponding quantum field that permeates all of space and time.
- **Field Quantization:** In QFT, particles are excitations or quanta of their respective fields. For example, the electron is a quantized excitation of the electron field, and photons are excitations of the electromagnetic field.

Interactions:

- **Force Mediation:** Fields interact with each other through the exchange of particles, known as force carriers or gauge bosons. For instance, the electromagnetic field interacts via the exchange of photons.
- **Lagrangian and Path Integrals:** The behavior of fields is described by a Lagrangian, which encapsulates the dynamics and interactions of the fields. Path integrals, a mathematical tool, are used to calculate the probabilities of different field configurations and interactions.

Strings in Quantum Field Theory

Strings as Extended Objects:

- **Vibrating Strings:** In string theory, the point-like particles of QFT are replaced by one-dimensional strings. The different vibrational modes of these strings correspond to different particles, including those found in the Standard Model of particle physics.
- **Open and Closed Strings:** Strings can be open, with two endpoints, or closed, forming a loop. Each type has different properties and roles in the theory.

From Particles to Strings:

- **Vibrational Modes:** Each vibrational mode of a string corresponds to a different particle. For example, a specific mode might correspond to an electron, while another mode might correspond to a photon or graviton.
- **Extended Dynamics:** Unlike point particles, strings interact by joining and splitting, a process that is naturally extended over a finite area, helping to avoid the infinities often encountered in point-particle theories.

Quantum Fields on the String Worldsheet

Worldsheet Dynamics:

- **Worldsheet:** The two-dimensional surface traced out by a string as it moves through spacetime is called a worldsheet. This is analogous to the trajectory of a point particle but extended into a two-dimensional surface.
- **Conformal Field Theory (CFT):** The dynamics of the fields on the string worldsheet are described by conformal field theory. CFT ensures that the physics is invariant under conformal transformations, preserving the angles but not necessarily the distances.

Vertex Operators:

- **Insertion Points:** Interactions on the worldsheet are represented by vertex operators, which insert specific states at points on the worldsheet. These operators help to describe how strings interact and transform into different vibrational modes.
- **Physical States:** The physical states of the strings are constructed using these vertex operators, ensuring that the resulting particles obey the necessary symmetries and constraints of the theory.

Unified Description of Forces

Gauge Fields and Strings:

- **Open Strings and Gauge Bosons:** The endpoints of open strings can carry charges and interact with gauge fields, which describe the fundamental forces. For example, the vibrations of open strings can correspond to the gauge bosons of the electromagnetic, weak, and strong forces.
- **Closed Strings and Gravity:** Closed strings inherently include the graviton, the quantum of the gravitational field, providing a natural framework for quantum gravity.

Superstring Theory:

- **Supersymmetry:** Superstring theory incorporates supersymmetry, a symmetry that pairs each boson with a corresponding fermion, and vice versa. This symmetry helps to resolve many theoretical inconsistencies and provides a more complete picture of fundamental interactions.
- **Ten Dimensions:** For consistency, superstring theory requires ten dimensions. The extra six dimensions are compactified into complex shapes, often modeled by Calabi-Yau manifolds, which influence the properties of the strings and the resulting particles.

Mathematical Techniques

Polyakov Action:

- **Action Principle:** The Polyakov action describes the dynamics of the string worldsheet, similar to how the action principle in classical mechanics describes the motion of point particles. This action is used to derive the equations of motion for strings.
- **Path Integrals:** In string theory, path integrals are extended to sum over all possible worldsheet configurations. This approach helps calculate the probabilities of various string interactions and their corresponding particle states.

Dualities:

- **T-Duality and S-Duality:** Dualities are mathematical transformations that reveal deep connections between different string theories. T-duality relates theories compactified on large and small radii, while S-duality connects strong and weak coupling regimes. These dualities show that different string theories are actually different aspects of a single underlying framework.

Physical Implications

Unification of Forces:

- **Single Framework:** String theory aims to provide a single framework that unifies all fundamental forces. The interactions between strings naturally give rise to the gauge bosons and gravitons, integrating the principles of quantum field theory with gravity.
- **Predictive Power:** The vibrational modes of strings predict a spectrum of particles, potentially including those not yet discovered. This predictive power makes string theory a compelling candidate for a Theory of Everything.

Cosmology and Black Holes:

- **Early Universe:** String theory has implications for the early universe, offering potential explanations for the Big Bang and cosmic inflation. The framework can help address questions about the initial conditions and evolution of the universe.
- **Black Hole Information:** String theory provides insights into the black hole information paradox, suggesting that information is not lost but rather encoded in the vibrations of strings. The AdS/CFT correspondence, a conjecture in string theory, proposes a relationship between a higher-dimensional gravitational theory and a lower-dimensional quantum field theory, offering new ways to understand black holes and quantum gravity.

Conformal Field Theory

Conformal Field Theory (CFT) is a crucial component in the framework of string theory. It plays a central role in understanding the dynamics of strings and their interactions. Here's an in-depth look at CFT and its significance in string theory.

Basic Concepts of Conformal Field Theory

Definition:

- **Conformal Symmetry:** Conformal Field Theory studies quantum field theories that are invariant under conformal transformations. These transformations preserve angles but not necessarily distances. This symmetry is particularly powerful because it significantly constrains the form of the theory.
- **Two-Dimensional CFT:** In string theory, the worldsheet swept out by a string as it moves through spacetime is a two-dimensional surface. The dynamics of fields on this worldsheet are described by a two-dimensional CFT.

Holomorphic and Anti-Holomorphic Components:

- **Complex Coordinates:** In two-dimensional CFT, it's convenient to use complex coordinates zzz and $z^-\bar{z}z^-$ for the worldsheet. The fields can be decomposed into holomorphic (depending on zzz) and anti-holomorphic (depending on $z^-\bar{z}z^-$) components.
- **Holomorphic Functions:** These functions are central to the structure of CFT and simplify many calculations by taking advantage of the properties of complex analysis.

Role in String Theory

Worldsheet Description:

- **Action Principle:** The action that describes the dynamics of the string's worldsheet is typically the Polyakov action. This action is invariant under conformal transformations, making CFT the natural language for string dynamics.
- **Virasoro Algebra:** The local conformal symmetry on the worldsheet is captured by the Virasoro algebra, an infinite-dimensional algebra that generates the conformal transformations. This algebra is fundamental to the structure of CFT in string theory.

Vertex Operators:

- **Insertion Points:** Interactions on the worldsheet are represented by vertex operators. These operators are inserted at specific points in the complex plane and describe the emission and absorption of strings.

- **Physical States:** The physical states of the string correspond to different vertex operators. These operators must satisfy specific conditions, like conformal invariance, to be part of the physical spectrum.

Mathematical Framework

Central Charge:

- **Conformal Anomaly:** The central charge ccc is a key quantity in CFT that measures the extent of the conformal anomaly. For a consistent string theory, the total central charge must cancel the conformal anomaly. For example, in bosonic string theory, the central charge is 26, while in superstring theory, it is 15.
- **Critical Dimensions:** The requirement for anomaly cancellation leads to the determination of the critical dimension of string theory. For example, bosonic string theory requires 26 dimensions, while superstring theory requires 10 dimensions.

Correlation Functions:

- **Operator Product Expansion (OPE):** The OPE is a powerful tool in CFT that describes how products of operators behave as their insertion points approach each other. It provides a systematic way to compute correlation functions.
- **Conformal Blocks:** These are the building blocks of correlation functions in CFT. They encode the contributions of different conformal families and simplify the calculation of physical quantities.

Applications in String Theory

Moduli Space:

- **Parameter Space:** The moduli space of a string compactification describes the different possible shapes and sizes of the compactified dimensions. CFT helps explore this space by providing tools to study the conformal field theories associated with different points in moduli space.
- **Physical Implications:** The structure of the moduli space determines the physical properties of the resulting four-dimensional theory, including particle masses and interaction strengths.

Dualities and Symmetries:

- **T-Duality and S-Duality:** CFT is important in understanding dualities in string theory. For example, T-duality relates string theories compactified on large and small radii, revealing that seemingly different theories can be equivalent.

- **Mirror Symmetry:** Mirror symmetry is a duality between pairs of Calabi-Yau manifolds that result in equivalent physical theories. CFT provides the mathematical framework to study these dualities and their implications.

AdS/CFT Correspondence:

- **Holographic Principle:** The AdS/CFT correspondence, proposed by Juan Maldacena, is a conjecture that relates a string theory in anti-de Sitter (AdS) space to a conformal field theory on its boundary. This duality provides a powerful tool for studying quantum gravity and strongly interacting quantum field theories.
- **Boundary Theory:** According to this correspondence, the dynamics of the bulk AdS space can be fully described by a CFT living on the boundary of that space. This principle has led to significant insights into black hole physics, quantum gravity, and gauge theory.

Current Research and Challenges

Mathematical Rigor:

- **Advanced Techniques:** Research in CFT involves advanced mathematical techniques from algebraic geometry, topology, and representation theory. Developing new methods to solve CFTs and understand their implications is an ongoing challenge.
- **Exact Solutions:** Finding exact solutions to specific CFTs and understanding their properties continues to be a major area of research. These solutions help in making precise predictions in string theory and other areas of theoretical physics.

Experimental Connections:

- **Indirect Evidence:** Direct experimental verification of CFT predictions in string theory is challenging. However, researchers look for indirect evidence through high-energy physics experiments, cosmological observations, and condensed matter systems.
- **Cross-Disciplinary Impact:** CFT has applications beyond string theory, including in condensed matter physics, statistical mechanics, and quantum information theory. These connections help to test and refine the theoretical framework of CFT.

Overall, Conformal Field Theory is an essential component of string theory, providing the mathematical structure to describe the dynamics of strings on their worldsheet. Its principles and techniques are important for understanding string interactions, exploring the moduli space, and studying dualities and the AdS/CFT correspondence.

The Holographic Principle

The holographic principle is a concept in theoretical physics suggesting that all the information contained within a volume of space can be described by a theory operating on the boundary of that space. This idea, proposed by Gerard 't Hooft and further developed by Leonard Susskind, posits that the universe can be viewed as a two-dimensional information structure "painted" on the cosmological horizon.

Key Aspects:

- **Black Hole Entropy:** The principle was inspired by the study of black holes. Stephen Hawking and Jacob Bekenstein showed that the entropy of a black hole is proportional to the area of its event horizon, not its volume. This finding suggested that the information content of a black hole could be encoded on its two-dimensional boundary.
- **AdS/CFT Correspondence:** Juan Maldacena's AdS/CFT correspondence is a concrete realization of the holographic principle. It states that a string theory in a five-dimensional anti-de Sitter (AdS) space is equivalent to a conformal field theory (CFT) on its four-dimensional boundary. This duality means that the gravitational dynamics in the AdS space can be described by the quantum field theory on the boundary, providing a powerful tool for studying quantum gravity.

Implications:

- **Quantum Gravity:** The holographic principle offers a potential framework for understanding quantum gravity, suggesting that gravity and spacetime may emerge from more fundamental, lower-dimensional quantum theories.
- **Information Paradox:** It provides insights into resolving the black hole information paradox, implying that information is not lost in black holes but encoded on their surface.

CHAPTER 6: DIMENSIONS AND THE MULTIVERSE

Understanding Higher Dimensions

In string theory, understanding higher dimensions is critical. These additional dimensions are not just theoretical constructs; they provide the necessary framework for the consistency and unification of fundamental forces. Here's a detailed exploration of higher dimensions in the context of string theory.

The Concept of Higher Dimensions
Beyond the Familiar:

- **Three Spatial Dimensions:** We live in a universe with three spatial dimensions: length, width, and height. Along with time, these make up the four-dimensional spacetime of our everyday experience.
- **Extra Dimensions:** String theory posits that there are more than these four dimensions. The simplest versions of string theory require ten dimensions (nine spatial and one temporal), while M-theory, an extension of string theory, requires eleven dimensions.

Why Extra Dimensions?

- **Mathematical Consistency:** The additional dimensions are necessary for the mathematical consistency of string theory. Without them, the equations governing string interactions would not work correctly.
- **Unification of Forces:** Extra dimensions allow for the unification of all fundamental forces, including gravity, within a single theoretical framework.

Visualizing Higher Dimensions
Compactification:

- **Curled-Up Dimensions:** The extra dimensions in string theory are compactified, meaning they are curled up into tiny shapes that are too small to observe directly. This concept was proposed by Theodor Kaluza and Oskar Klein in the early 20th century.
- **Calabi-Yau Manifolds:** These compactified dimensions are often modeled using complex geometric shapes known as Calabi-Yau manifolds. These manifolds have specific properties that allow for the preservation of supersymmetry and influence the vibrational modes of strings.

Imagining Compactified Dimensions:

- **Garden Hose Analogy:** Imagine a garden hose. From a distance, it looks like a one-dimensional line. But up close, you can see it has a second dimension, the circular cross-section. Similarly, the compactified dimensions are tiny and hidden from our everyday perception.

Implications for Physics

Particle Properties:

- **Vibrational Modes:** The shape and size of the compactified dimensions determine the vibrational modes of strings. Different vibrational modes correspond to different particles, including those in the Standard Model of particle physics.
- **Mass and Charge:** The properties of particles, such as mass and charge, are influenced by how strings vibrate in these higher dimensions.

Force Unification:

- **Gauge Bosons:** The compactified dimensions help explain the existence of gauge bosons, the force carriers in particle physics. The interactions between strings and the geometry of the extra dimensions lead to the emergence of these particles.
- **Gravitons:** Gravity, which is much weaker than the other fundamental forces, can be understood as a string vibration in higher dimensions, with gravitons mediating this force.

Experimental Considerations

Detecting Extra Dimensions:

- **Indirect Evidence:** Direct detection of extra dimensions is beyond current experimental capabilities due to their small size. However, physicists look for indirect evidence through high-energy particle collisions and cosmological observations.
- **LHC and Beyond:** Experiments at the Large Hadron Collider (LHC) and future particle accelerators may provide hints about the existence of extra dimensions by observing deviations from the Standard Model predictions.

Cosmological Implications:

- **Early Universe:** The structure of higher dimensions could have influenced the early universe, affecting the formation of cosmic structures and the behavior of fundamental forces.
- **Dark Matter and Energy:** Some theories suggest that dark matter and dark energy could be manifestations of interactions involving extra dimensions.

The Multiverse Connection

Landscape of Solutions:

- **String Theory Landscape:** The vast number of possible shapes for compactified dimensions leads to a multitude of possible solutions, known as the string theory landscape. Each solution corresponds to a different possible universe with its own physical laws.
- **Multiverse Hypothesis:** This landscape suggests the existence of a multiverse, where our universe is just one of many, each with different properties determined by the configuration of extra dimensions.

Anthropic Principle:

- **Fine-Tuning:** The anthropic principle posits that the physical constants in our universe are finely tuned to allow for the existence of life. The multiverse hypothesis provides a potential explanation, suggesting that we live in one of the few universes where the conditions are right for life to develop.

Higher dimensions are a fundamental aspect of string theory, essential for its mathematical consistency and the unification of fundamental forces. These dimensions, though hidden from direct observation, shape the properties of particles and the nature of our universe. Understanding them opens new avenues for exploring the fundamental nature of reality and our place within a possible multiverse.

String Theory's Extra Dimensions

String theory proposes that the universe contains more than the familiar three spatial dimensions and one-time dimension. Specifically, string theory requires ten dimensions: nine spatial and one temporal. These extra dimensions are essential for the mathematical consistency of the theory and for the unification of all fundamental forces.

Compactification

Curled-Up Dimensions:

- **Hidden Dimensions:** The extra dimensions are compactified, meaning they are curled up into tiny, unobservable shapes. This idea was first introduced by Theodor Kaluza and Oskar Klein in the 1920s to unify electromagnetism and gravity.
- **Calabi-Yau Manifolds:** These compactified dimensions are often modeled using complex geometric shapes known as Calabi-Yau manifolds.

These shapes ensure the preservation of supersymmetry, a crucial feature in string theory.

Implications

Particle Properties:

- **Vibrational Modes:** The different vibrational modes of strings in these compactified dimensions correspond to different particles. The properties of these particles, such as mass and charge, depend on the shape and size of the extra dimensions.
- **Force Carriers:** The interactions of strings in higher dimensions can explain the existence of gauge bosons and gravitons, which mediate fundamental forces.

Experimental Search

Indirect Evidence:

- **Particle Collisions:** While direct observation of extra dimensions is not currently possible, physicists search for indirect evidence in high-energy particle collisions, such as those at the Large Hadron Collider (LHC).
- **Cosmological Observations:** Observations of the early universe and cosmic phenomena may provide clues about the influence of extra dimensions.

Multiverse Hypothesis

String Theory Landscape:

- **Multiple Solutions:** The vast number of possible configurations of the compactified dimensions leads to a multitude of solutions, known as the string theory landscape.
- **Different Universes:** Each configuration corresponds to a different universe, potentially leading to the idea of a multiverse where different physical laws and constants apply.

Compactification

Compactification is a key concept in string theory that addresses how extra dimensions, required by the theory, become unobservable in our everyday experience. While string theory posits that the universe has more than the familiar four dimensions (three spatial and one temporal), the additional dimensions are compactified, or curled up, into incredibly small scales. Here's a detailed look at compactification and its implications.

The Need for Extra Dimensions

Mathematical Consistency:

- **String Theory's Requirements:** For string theory to be mathematically consistent, it requires ten dimensions (nine spatial and one temporal). M-theory, an extension of string theory, requires eleven dimensions.
- **Unifying Forces:** These extra dimensions allow string theory to unify all fundamental forces, including gravity, into a single theoretical framework.

What is Compactification?

Curled-Up Dimensions:

- **Invisible to the Naked Eye:** Compactification refers to the process where the extra dimensions are curled up into shapes so tiny that they are undetectable at everyday scales. This concept was first introduced by Theodor Kaluza and Oskar Klein in the early 20th century.
- **Calabi-Yau Manifolds:** The shapes of these compactified dimensions are often modeled using Calabi-Yau manifolds. These are complex, multi-dimensional geometric structures that have specific properties, such as being Ricci-flat, which preserve supersymmetry.

Visualizing Compactified Dimensions

Analogies to Understand Compactification:

- **Garden Hose Analogy:** Imagine a garden hose. From a distance, it appears as a one-dimensional line. Up close, you can see its circular cross-section, revealing a hidden dimension. Similarly, the extra dimensions in string theory are compactified to such a small scale that they are not apparent in everyday observations.

Implications of Compactification

Particle Properties:

- **Vibrational Modes:** The different vibrational modes of strings in the compactified dimensions correspond to different particles. The specific properties of these particles, such as mass and charge, depend on the shape and size of the compactified dimensions.
- **Force Carriers:** The compactified dimensions help explain the existence of gauge bosons (which mediate the electromagnetic, weak, and strong forces) and gravitons (which mediate gravity).

Moduli Space:

- **Parameter Space:** The moduli space describes all possible shapes and sizes of the compactified dimensions. Each point in this space represents a different compactification scenario, influencing the physical properties of the resulting four-dimensional universe.
- **Physical Constants:** The values of physical constants, like particle masses and coupling constants, are determined by the position within the moduli space. Different compactifications can lead to different physical laws.

Searching for Evidence

Indirect Detection:

- **High-Energy Experiments:** While directly observing the compactified dimensions is not possible with current technology, physicists search for indirect evidence through high-energy particle collisions, such as those conducted at the Large Hadron Collider (LHC). Deviations from the Standard Model predictions could hint at the presence of extra dimensions.
- **Cosmological Observations:** The structure of the early universe and cosmic microwave background radiation may provide clues about the influence of extra dimensions. Certain patterns and anomalies could suggest the effects of these hidden dimensions.

Connection to the Multiverse

String Theory Landscape:

- **Multiple Solutions:** The compactification process leads to a vast number of possible shapes for the extra dimensions, known as the string theory landscape. Each compactification corresponds to a different vacuum state with its own set of physical laws.
- **Multiverse Hypothesis:** This landscape suggests the existence of a multiverse, where each universe has a unique configuration of compactified dimensions and distinct physical laws. Our universe is just one of many in this vast multiverse.

Anthropic Principle:

- **Fine-Tuning:** The anthropic principle suggests that the physical constants in our universe are finely tuned to allow for the existence of life. The multiverse hypothesis provides a potential explanation: we observe these constants because we live in one of the few universes where conditions are suitable for life.

Challenges and Future Directions

Mathematical Complexity:

- **Advanced Techniques:** Understanding and solving the equations governing compactified dimensions requires advanced mathematical techniques from algebraic geometry, topology, and differential geometry. Ongoing research aims to develop these tools further and explore the implications of different compactifications.

Experimental Advances:

- **Technological Progress:** Future advancements in technology and experimental methods may provide more direct evidence of extra dimensions. High-energy physics experiments, precision measurements, and new observational techniques in cosmology could offer deeper insights.

Compactification is a fundamental process in string theory that hides extra dimensions by curling them up into tiny, complex shapes. This concept is important for the theory's mathematical consistency and for explaining the properties of particles and forces in our four-dimensional universe. Understanding compactification opens new avenues for exploring the fundamental nature of reality and the potential existence of a multiverse.

The Concept of the Multiverse

The concept of the multiverse posits that our universe is just one of many that exist. These universes, collectively referred to as the multiverse, each have their own distinct laws of physics, constants, and even different dimensions. This idea, while still speculative, is rooted in several areas of theoretical physics and cosmology, including string theory and inflationary cosmology. Here's a detailed look at the concept of the multiverse.

Origins of the Multiverse Concept

Quantum Mechanics:

- **Many-Worlds Interpretation:** One of the earliest hints at a multiverse comes from the many-worlds interpretation of quantum mechanics, proposed by Hugh Everett in 1957. This interpretation suggests that all possible outcomes of quantum events actually occur, each in a separate, branching universe.

Cosmology:

- **Inflationary Theory:** The theory of cosmic inflation, proposed by Alan Guth in the 1980s, posits that the early universe underwent a rapid expansion. This expansion could produce an infinite number of bubble universes, each with its own physical properties.

- **Eternal Inflation:** Further developments in inflationary theory, particularly the concept of eternal inflation, suggest that inflation continues in some regions of space, constantly creating new bubble universes.

Multiverse in String Theory

String Theory Landscape:

- **Multiple Vacua:** String theory predicts a vast number of possible vacuum states, each corresponding to a different configuration of compactified dimensions. This multitude of solutions is known as the string theory landscape.
- **Different Physical Laws:** Each vacuum state could give rise to a universe with different physical constants and laws of physics, explaining why our universe has the specific properties that allow for life.

Types of Multiverses

Level I: Beyond Our Cosmic Horizon:

- **Observable vs. Unobservable Universe:** The simplest type of multiverse is one where regions of space beyond our observable universe exist. These regions follow the same physical laws but differ in initial conditions.

Level II: Bubble Universes:

- **Inflationary Multiverse:** In eternal inflation, different regions of space stop inflating at different times, creating bubble universes. These bubbles form separate universes, each with its own physical properties.
- **Different Constants:** Each bubble universe might have different values for fundamental constants, resulting in diverse physical realities.

Level III: Many-Worlds Interpretation:

- **Quantum Branching:** According to the many-worlds interpretation of quantum mechanics, every quantum event spawns multiple branching universes, each representing a different outcome of the event.

Level IV: Mathematical Universes:

- **Ultimate Ensemble:** This idea, proposed by Max Tegmark, suggests that all possible mathematical structures correspond to physical realities. Each mathematical structure forms its own universe with its own laws of physics.

Implications of the Multiverse

Anthropic Principle:

- **Fine-Tuning:** The anthropic principle, as we covered in a separate section, posits that we observe certain physical constants because they allow for the existence of life. In a multiverse, regions with different constants exist, but only those with life-permitting values are observed by conscious beings.

Challenges to Verification:

- **Empirical Evidence:** One of the major challenges of the multiverse concept is the lack of direct empirical evidence. Observing or interacting with other universes is beyond our current technological capabilities.
- **Indirect Evidence:** Scientists look for indirect evidence, such as cosmic microwave background radiation patterns or high-energy physics experiments, that might hint at the existence of other universes.

Philosophical and Scientific Considerations

Philosophical Debates:

- **Nature of Reality:** The multiverse concept raises profound questions about the nature of reality, existence, and what it means to be a universe.
- **Scientific Validity:** Some critics argue that the multiverse is not a scientifically valid theory because it is difficult, if not impossible, to test or falsify.

Scientific Benefits:

- **Explaining Fine-Tuning:** The multiverse provides a potential explanation for the fine-tuning of the constants of nature, which seem improbably suited for life.
- **Advancing Theoretical Physics:** Exploring the implications of the multiverse drives advancements in theoretical physics, cosmology, and our understanding of the fundamental nature of the universe.

Current Research and Future Directions

Theoretical Developments:

- **String Theory and Quantum Gravity:** Researchers continue to explore the implications of string theory and quantum gravity for the multiverse concept, seeking to develop a more coherent theoretical framework.
- **Mathematical Models:** Advanced mathematical models are being developed to better understand the properties and behaviors of different possible universes within the multiverse.

Experimental Prospects:

- **Cosmological Observations:** Future observations of the cosmic microwave background, large-scale structure, and other cosmological phenomena may provide indirect evidence supporting the multiverse.
- **High-Energy Physics:** Experiments at particle accelerators, such as the Large Hadron Collider, might reveal new particles or interactions hinting at the existence of other dimensions or universes.

The concept of the multiverse proposes the existence of multiple, possibly infinite, universes each with its own unique properties. This idea emerges from various theories in physics and cosmology, providing potential explanations for the fine-tuning of physical constants and the nature of reality. While challenging to test, the multiverse remains a compelling and influential idea in modern theoretical physics.

CHAPTER 7: STRING THEORY AND RELATIVITY

General Relativity Overview

General relativity, formulated by Albert Einstein in 1915, revolutionized our understanding of gravity. Unlike Newton's theory, which described gravity as a force between masses, general relativity describes gravity as the curvature of spacetime caused by mass and energy. Let's cover an overview of the key concepts in general relativity.

The Principle of Equivalence
Foundational Idea:

- **Equivalence of Inertial and Gravitational Mass:** Einstein's principle of equivalence states that the effects of gravity are indistinguishable from the effects of acceleration. For example, standing in an accelerating elevator feels the same as standing on Earth due to gravity.

Spacetime and Curvature
Spacetime Concept:

- **Four-Dimensional Fabric:** General relativity combines the three dimensions of space with the one dimension of time into a four-dimensional continuum called spacetime.
- **Curvature by Mass and Energy:** Massive objects like stars and planets warp the spacetime around them. This curvature is what we perceive as gravity.

Metric Tensor:

- **Describing Spacetime:** The metric tensor is a mathematical object that describes the geometry of spacetime. It tells us how distances and times are measured in curved spacetime.
- **Einstein's Field Equations:** These equations relate the curvature of spacetime (described by the Einstein tensor) to the distribution of mass and energy (described by the stress-energy tensor). The field equations are written as $G\mu\nu = 8\pi T\mu\nu$, where $G\mu\nu$ is the Einstein tensor and $T\mu\nu$ is the stress-energy tensor.

Geodesics and Free Fall
Path of Least Resistance:

- **Geodesics:** In general relativity, free-falling objects move along paths called geodesics, which are the straightest possible paths in curved spacetime. These paths minimize the spacetime interval, similar to how a straight line is the shortest path between two points in flat space.

Gravity as Curvature:

- **No Force:** Unlike in Newtonian mechanics, gravity is not considered a force in general relativity. Instead, objects follow geodesics in curved spacetime, and the apparent force of gravity is a result of this curvature.

Predictions and Confirmations

Bending of Light:

- **Gravitational Lensing:** Light from distant stars bends when it passes near massive objects, an effect known as gravitational lensing. This phenomenon was first observed during a solar eclipse in 1919, providing early confirmation of general relativity.

Perihelion Precession:

- **Mercury's Orbit:** The orbit of Mercury precesses slightly over time, an effect not fully explained by Newtonian mechanics. General relativity accounts for this discrepancy, accurately predicting the observed precession.

Gravitational Waves:

- **Ripples in Spacetime:** General relativity predicts the existence of gravitational waves, which are ripples in spacetime caused by accelerating massive objects, such as merging black holes. These waves were first directly detected by LIGO in 2015, confirming another key prediction of the theory.

Black Holes and Singularities

Extreme Curvature:

- **Black Holes:** When a massive star collapses, it can form a black hole, an object with gravity so strong that not even light can escape. The boundary of a black hole is called the event horizon.
- **Singularities:** At the center of a black hole lies a singularity, a point where spacetime curvature becomes infinite, and the laws of physics as we know them break down.

Cosmology and General Relativity

Expanding Universe:

- **Cosmological Models:** General relativity provides the framework for modern cosmology. It describes the dynamics of an expanding universe, as first evidenced by Edwin Hubble's observation of redshifts in distant galaxies.
- **Cosmological Constant:** Einstein introduced the cosmological constant, $\Lambda\backslash Lambda\Lambda$, to allow for a static universe. Although he later discarded it, the concept has resurfaced in the context of dark energy and the accelerating expansion of the universe.

Integration with Quantum Mechanics

Need for Quantum Gravity:

- **Incompatibility:** General relativity is a classical theory and does not include quantum effects. This incompatibility with quantum mechanics becomes significant in extreme conditions, such as near black holes or during the Big Bang.
- **String Theory's Role:** String theory aims to unify general relativity with quantum mechanics, providing a framework for quantum gravity. It posits that the fundamental constituents of the universe are one-dimensional strings, whose vibrations correspond to different particles, including the graviton, the hypothetical quantum particle that mediates gravity.

General relativity provides a comprehensive description of gravity as the curvature of spacetime caused by mass and energy. Its predictions have been confirmed by numerous experiments and observations, profoundly influencing our understanding of the universe. Integrating general relativity with quantum mechanics remains one of the foremost challenges in theoretical physics, with string theory offering a promising approach.

String Theory's Compatibility with Relativity

String theory is a theoretical framework that aims to unify all fundamental forces and particles within a single coherent model. One of its key strengths is its compatibility with Einstein's theory of general relativity, which describes gravity. Here's an in-depth look at how string theory aligns with and extends the principles of relativity.

The Challenge of Unification

Classical and Quantum Divide:

- **General Relativity:** Describes gravity as the curvature of spacetime caused by mass and energy. It is a classical theory that works exceptionally well on large scales, such as stars and galaxies.
- **Quantum Mechanics:** Governs the behavior of particles at the smallest scales. It includes three of the four fundamental forces—electromagnetism, the weak nuclear force, and the strong nuclear force. These forces are described by quantum field theories.

Incompatibility Issue:

- **Singularities and Infinities:** When trying to apply quantum principles to gravity, the equations often produce infinities, such as those found in black holes and the Big Bang. This incompatibility indicates the need for a new theoretical framework.

How String Theory Resolves This

Strings as Fundamental Entities:

- **One-Dimensional Strings:** In string theory, the fundamental objects are not point-like particles but one-dimensional strings. These strings can vibrate at different frequencies, with each vibrational mode corresponding to a different particle.
- **Avoiding Infinities:** The extended nature of strings helps smooth out the infinities that arise in point-particle theories. This smoothing effect is crucial for making sense of gravity at the quantum level.

Incorporation of Gravitons:

- **Graviton Emergence:** One of the vibrational modes of a closed string corresponds to the graviton, a hypothetical quantum particle that mediates the force of gravity. The existence of the graviton within string theory naturally incorporates quantum gravity into the framework.
- **Gravity as Curvature:** String theory retains the concept of gravity as the curvature of spacetime, consistent with general relativity, but extends this concept to include quantum mechanical effects.

Mathematical Framework

Higher Dimensions:

- **Extra Dimensions:** String theory requires additional spatial dimensions for consistency—ten in superstring theory and eleven in M-theory. These extra dimensions are compactified, meaning they are curled up into small, unobservable shapes.

- **Calabi-Yau Manifolds:** The compactified dimensions are often modeled using complex geometric shapes known as Calabi-Yau manifolds. These shapes influence the vibrational modes of strings and, consequently, the properties of particles and forces.

Conformal Field Theory:

- **Worldsheet Dynamics:** The dynamics of strings are described by conformal field theory (CFT) on the two-dimensional surface traced out by the string, known as the worldsheet. CFT ensures the theory is invariant under conformal transformations, providing a consistent framework for string interactions.

Dualities and Unified Descriptions
String Dualities:

- **T-Duality:** Relates string theories compactified on large and small radii, showing that these theories are equivalent descriptions of the same physics.
- **S-Duality:** Connects strong and weak coupling regimes of string theories, indicating that a strongly interacting string theory can be equivalent to a weakly interacting one. These dualities reveal that different string theories are different aspects of a single underlying theory.

AdS/CFT Correspondence:

- **Holographic Principle:** The AdS/CFT correspondence, proposed by Juan Maldacena, is a duality between a string theory formulated in a higher-dimensional anti-de Sitter (AdS) space and a conformal field theory on its lower-dimensional boundary. This correspondence provides a powerful tool for studying quantum gravity and has profound implications for our understanding of spacetime and black holes.

Physical Implications
Black Holes and Singularities:

- **Resolution of Singularities:** String theory suggests that singularities, such as those at the centers of black holes, are smoothed out by the extended nature of strings. This offers a potential resolution to the infinite curvature predicted by general relativity.
- **Hawking Radiation and Information Paradox:** String theory provides insights into black hole thermodynamics and the information paradox, suggesting that information is not lost in black holes but encoded in subtle ways, possibly through holographic principles.

Early Universe and Cosmology:

- **Inflation and Structure Formation:** String theory has implications for the early universe, including mechanisms for cosmic inflation and the formation of large-scale structures. The theory's extra dimensions and compactification scenarios can influence the dynamics of the early universe.
- **Dark Matter and Dark Energy:** String theory offers potential explanations for dark matter and dark energy, proposing new particles and fields that could account for these mysterious components of the universe.

Experimental Prospects

High-Energy Physics:

- **LHC and Beyond:** While direct evidence for string theory is challenging to obtain due to the small scale of strings, high-energy experiments like those at the Large Hadron Collider (LHC) search for indirect evidence, such as deviations from the Standard Model predictions and new particle discoveries.

Cosmological Observations:

- **Cosmic Microwave Background:** Observations of the cosmic microwave background and large-scale structure of the universe may provide clues about the compactification of extra dimensions and the influence of string theory on cosmic evolution.

Bridging Quantum Mechanics and Relativity

Bridging quantum mechanics and general relativity is one of the most profound challenges in theoretical physics. Quantum mechanics describes the behavior of particles at the smallest scales, while general relativity explains the force of gravity and the structure of spacetime on large scales. These two pillars of modern physics operate on fundamentally different principles and have resisted unification for decades.

Quantum Mechanics: The Micro World

Principles of Quantum Mechanics:

- **Wave-Particle Duality:** Particles exhibit both wave-like and particle-like properties. The behavior of particles is described by wavefunctions, which provide probabilities for finding a particle in a particular state.

- **Uncertainty Principle:** Heisenberg's Uncertainty Principle states that one cannot simultaneously know the exact position and momentum of a particle. This introduces fundamental limits to measurement.
- **Quantization:** Physical quantities such as energy are quantized, meaning they can only take on discrete values. This is evident in the energy levels of electrons in atoms.

Mathematical Framework:

- **Schrödinger Equation:** Describes how the quantum state of a physical system changes over time.
- **Path Integrals:** Proposed by Richard Feynman, this approach sums over all possible paths that a particle can take to compute probabilities of different outcomes.

General Relativity: The Macro World

Principles of General Relativity:

- **Spacetime Curvature:** Gravity is not a force between masses but a result of the curvature of spacetime caused by mass and energy. Objects follow geodesics, the straightest possible paths in this curved spacetime.
- **Equivalence Principle:** The effects of gravity are locally indistinguishable from acceleration. This principle led Einstein to generalize his theory of relativity.
- **Field Equations:** Einstein's field equations relate the curvature of spacetime (described by the Einstein tensor) to the energy and momentum within that spacetime (described by the stress-energy tensor).

The Conflict

Incompatibilities:

- **Scale Differences:** Quantum mechanics operates at microscopic scales, while general relativity governs large-scale structures. Their principles do not seamlessly integrate, especially in extreme conditions like black holes or the Big Bang.
- **Singularities:** General relativity predicts singularities, where densities become infinite and the laws of physics break down. Quantum mechanics struggles to describe these singularities.
- **Non-Renormalizability:** When attempting to apply quantum field theory techniques to gravity, the calculations result in non-renormalizable infinities, making it impossible to make finite predictions.

String Theory's Approach

Strings Replace Particles:

- **One-Dimensional Strings:** In string theory, the fundamental constituents of the universe are not point particles but one-dimensional strings. These strings can vibrate at different frequencies, with each mode corresponding to a different particle.
- **Finite Size:** The finite size of strings helps smooth out the infinities that arise in point-particle theories, providing a potential solution to the problem of singularities.

Incorporating Gravity:

- **Graviton:** A specific vibrational mode of a closed string corresponds to the graviton, the quantum particle that mediates the force of gravity. This naturally incorporates gravity into the quantum framework.
- **Extra Dimensions:** String theory requires additional spatial dimensions for mathematical consistency—ten in superstring theory and eleven in M-theory. These extra dimensions are compactified into small, complex shapes.

Mathematical Consistency:

- **Supersymmetry:** String theory often incorporates supersymmetry, a theoretical symmetry that pairs each particle with a superpartner. This helps resolve various theoretical issues and provides a more coherent framework.
- **Conformal Field Theory:** The dynamics of strings are described using conformal field theory on the worldsheet, ensuring that the physics is invariant under conformal transformations.

Unifying Principles

Quantum Gravity:

- **Path Integrals:** In string theory, path integrals sum over all possible configurations of the string worldsheet, extending the quantum mechanical path integral approach to a higher-dimensional framework.
- **Dualities:** String theory reveals deep connections between seemingly different theories through dualities, such as T-duality and S-duality, showing that different string theories are actually different aspects of the same underlying theory.

AdS/CFT Correspondence:

- **Holographic Principle:** Proposed by Juan Maldacena, the AdS/CFT correspondence posits a duality between a string theory formulated in a higher-dimensional anti-de Sitter (AdS) space and a conformal field theory on its boundary. This duality provides a powerful tool for studying quantum gravity and offers insights into the nature of spacetime and black holes.

Experimental Prospects

Searching for Evidence:

- **High-Energy Physics:** Experiments at particle accelerators, such as the Large Hadron Collider (LHC), search for indirect evidence of string theory, such as the discovery of superpartners or deviations from the Standard Model.
- **Cosmology:** Observations of the cosmic microwave background, large-scale structure, and gravitational waves provide clues about the early universe and the potential effects of extra dimensions.

Theoretical Developments:

- **Mathematical Rigor:** Ongoing research in string theory involves developing advanced mathematical techniques from algebraic geometry, topology, and conformal field theory to explore the theory's implications.
- **Interdisciplinary Impact:** Insights from string theory contribute to various fields, including condensed matter physics, quantum information theory, and cosmology, highlighting its broad impact on our understanding of the universe.

Gravitons in String Theory

In string theory, the graviton is a fundamental particle that mediates the force of gravity. Unlike in traditional quantum field theories where particles are point-like, in string theory, all particles, including the graviton, are manifestations of one-dimensional strings vibrating at specific frequencies. Here's a detailed look at gravitons in the context of string theory.

Gravitons: The Basics

Definition:

- **Quantum of Gravity:** The graviton is the hypothetical quantum particle that carries the gravitational force, analogous to how the photon carries the electromagnetic force. It is a massless, spin-2 boson, meaning it has two units of intrinsic angular momentum.

Gravitons in String Theory

Vibrational Modes:

- **Closed Strings:** In string theory, the graviton arises as a specific vibrational mode of a closed string. Closed strings are loops without endpoints, and their vibrational patterns can give rise to different particles, depending on the mode of vibration.
- **Massless State:** The graviton corresponds to a massless vibrational state of the closed string, characterized by a symmetric, traceless tensor field, which matches the properties required for a spin-2 particle.

Mathematical Framework:

- **Polyakov Action:** The dynamics of strings, including closed strings, are described by the Polyakov action. This action is invariant under conformal transformations, ensuring consistency with the principles of quantum mechanics and general relativity.
- **Worldsheet Theory:** The worldsheet of a closed string, the two-dimensional surface traced out as it moves through spacetime, is governed by conformal field theory (CFT). The properties of the graviton are derived from the solutions to the equations of motion on this worldsheet.

Incorporating Gravity
Unification of Forces:

- **Natural Inclusion:** Unlike in quantum field theory, where gravity is difficult to incorporate due to issues with renormalizability, string theory naturally includes gravity through the existence of the graviton as a vibrational mode of the string.
- **Extra Dimensions:** For mathematical consistency, string theory requires additional spatial dimensions. In superstring theory, there are ten dimensions, while M-theory requires eleven. These extra dimensions are compactified into small, complex shapes known as Calabi-Yau manifolds.

Avoiding Infinities:

- **Smoothing Out Singularities:** The extended nature of strings helps smooth out the infinities and singularities that arise in point-particle theories. This property is crucial for making sense of gravity at the quantum level and resolving issues like black hole singularities.

Gravitons and Quantum Gravity
Path Integrals:

- **Summing Over Histories:** In string theory, path integrals extend to summing over all possible configurations of the string worldsheet. This approach helps calculate the probabilities of various string interactions, including those involving gravitons.
- **Graviton Exchange:** In the low-energy limit, the exchange of gravitons between massive objects results in the familiar inverse-square law of gravity described by Newton's law and general relativity.

AdS/CFT Correspondence:

- **Holographic Principle:** The AdS/CFT correspondence, a conjecture in string theory, posits a duality between a gravitational theory in a higher-dimensional anti-de Sitter (AdS) space and a conformal field theory on its boundary. This duality provides a powerful framework for studying quantum gravity and the behavior of gravitons.
- **Graviton Behavior:** In this context, the behavior of gravitons in the bulk AdS space can be understood in terms of the dynamics of the boundary CFT, offering insights into the quantum nature of gravity.

Experimental Prospects

Indirect Evidence:

- **High-Energy Experiments:** Direct detection of gravitons is extremely challenging due to their weak interaction with matter. However, high-energy physics experiments, such as those conducted at the Large Hadron Collider (LHC), search for indirect signs of string theory, such as deviations from the Standard Model predictions or evidence of extra dimensions.
- **Cosmological Observations:** Observations of gravitational waves, cosmic microwave background radiation, and large-scale structure of the universe may provide indirect evidence supporting the existence of gravitons and the validity of string theory.

Theoretical Developments:

- **Mathematical Rigor:** Ongoing research in string theory involves developing advanced mathematical techniques to better understand the properties and interactions of gravitons. This includes exploring dualities, compactification scenarios, and the implications of the holographic principle.
- **Interdisciplinary Impact:** Insights from the study of gravitons in string theory contribute to various fields, including cosmology, black hole physics, and quantum information theory, highlighting the broad impact of string theory on our understanding of the universe.

Challenges and Future Directions

Unresolved Issues:

- **Testing Predictions:** One of the significant challenges for string theory and the concept of gravitons is finding ways to test its predictions experimentally. The energy scales at which string effects become apparent are typically much higher than what current technology can probe.
- **Mathematical Complexity:** The mathematics of string theory, particularly in relation to gravitons and quantum gravity, is highly complex. Further advancements in theoretical and mathematical physics are necessary to fully realize and test the theory's potential.

In summary, in string theory, gravitons are the quantum particles that mediate gravity, arising as specific vibrational modes of closed strings. This natural incorporation of gravity, along with the unifying framework provided by string theory, offers a promising path toward understanding quantum gravity and bridging the gap between quantum mechanics and general relativity.

CHAPTER 8: TYPES OF STRING THEORY

Bosonic String Theory

Bosonic string theory is the earliest and simplest version of string theory, laying the foundation for later, more complex models. It was developed in the late 1960s and early 1970s as physicists sought to understand the fundamental forces and particles of nature.

Basics of Bosonic String Theory
One-Dimensional Strings:

- **Fundamental Entities:** In bosonic string theory, the fundamental objects are one-dimensional strings rather than point particles. These strings can vibrate in different modes, and each mode corresponds to a different particle.
- **Types of Strings:** Strings can be open, with endpoints, or closed, forming loops. Both types have different vibrational patterns and physical implications.

Vibrational Modes:

- **Particle Spectrum:** The various vibrational modes of a string determine the type of particle it represents. For example, specific vibrational patterns can correspond to particles like the photon or graviton.
- **Only Bosons:** As the name suggests, bosonic string theory includes only bosons—particles that carry forces. It does not include fermions, which make up matter. This limitation is significant because it prevents bosonic string theory from describing all known particles.

Mathematical Framework
Action and Worldsheet:

- **Polyakov Action:** The dynamics of strings are described by the Polyakov action, which governs how the worldsheet (the two-dimensional surface traced out by the string) evolves in spacetime.
- **Conformal Symmetry:** The theory is invariant under conformal transformations, meaning the equations describing the string's motion remain unchanged under specific changes in the shape of the worldsheet.

Critical Dimension:

- **26 Dimensions:** For mathematical consistency, bosonic string theory requires spacetime to have 26 dimensions. Of these, 22 must be compactified into very small, unseen dimensions, leaving the familiar four dimensions (three spatial and one temporal) visible.

Key Features and Challenges

Tachyons:

- **Instability:** Bosonic string theory predicts the existence of a tachyon, a hypothetical particle with imaginary mass that implies the theory is unstable. This instability indicates that the theory's vacuum state is not the true ground state, making it problematic for describing the real world.

No Supersymmetry:

- **Lack of Fermions:** Unlike superstring theories, bosonic string theory does not include supersymmetry, a symmetry that relates bosons and fermions. Without supersymmetry, the theory cannot incorporate fermions, which are essential for describing matter particles like electrons and quarks.

Historical Significance and Development

Foundation for Future Theories:

- **Pioneering Model:** Despite its limitations, bosonic string theory was groundbreaking as the first model to propose strings as the fundamental building blocks of the universe. It set the stage for the development of superstring theories, which incorporate both bosons and fermions and address many of the issues present in bosonic string theory.

Evolution to Superstring Theory:

- **Inclusion of Supersymmetry:** To overcome the limitations of bosonic string theory, physicists developed superstring theories in the 1980s. These theories incorporate supersymmetry, allowing for a consistent inclusion of both bosons and fermions, and require only 10 dimensions instead of 26.

Superstring Theory

Superstring theory is a refined version of string theory that addresses many limitations of the original bosonic string theory. Developed in the 1980s, it incorporates supersymmetry, which pairs each boson (force-carrying particle) with a fermion (matter particle), providing a more complete and realistic framework.

Basics of Superstring Theory

One-Dimensional Strings:

- **Fundamental Entities:** Just like in bosonic string theory, the fundamental objects are one-dimensional strings. These strings can vibrate in different modes, with each mode corresponding to different particles.
- **Types of Strings:** Strings can be open, with endpoints, or closed, forming loops.

Supersymmetry:

- **Bosons and Fermions:** Supersymmetry is a critical feature of superstring theory. It posits that every boson has a corresponding fermion and vice versa. This symmetry helps to balance the types of particles in the theory, allowing it to describe both matter and forces.
- **Consistency:** Supersymmetry helps cancel out certain mathematical infinities, making the theory more consistent and stable.

Mathematical Framework

Critical Dimensions:

- **Ten Dimensions:** Unlike bosonic string theory's 26 dimensions, superstring theory requires ten dimensions for mathematical consistency. Six of these dimensions are compactified into tiny, complex shapes, often modeled as Calabi-Yau manifolds, leaving four observable dimensions.

Five Versions:

- **Types I, IIA, IIB, HO, and HE:** There are five consistent versions of superstring theory, each with different characteristics regarding the types of strings and symmetries involved. Despite these differences, they are believed to be different aspects of a single, underlying theory known as M-theory.

Key Features and Advances

Incorporating Gravity:

- **Graviton:** Superstring theory naturally includes gravity. One of the vibrational modes of a closed string corresponds to the graviton, the hypothetical quantum particle that mediates the gravitational force.
- **Quantum Gravity:** This inclusion offers a framework for understanding quantum gravity, a significant step toward unifying general relativity and quantum mechanics.

No Tachyons:

- **Stability:** Unlike bosonic string theory, superstring theory does not predict tachyons, thus avoiding the instability problem.

Implications and Future Directions

Unification of Forces:

- **Theory of Everything:** Superstring theory aims to unify all fundamental forces—gravity, electromagnetism, and the nuclear forces—within a single theoretical framework.

Research and Experimentation:

- **Experimental Evidence:** Direct evidence for superstring theory is challenging to obtain, but researchers look for indirect signs, such as supersymmetric particles, in high-energy experiments like those at the Large Hadron Collider (LHC).

M-Theory

M-theory is a unifying framework in theoretical physics that seeks to reconcile and extend the five different versions of superstring theory. Proposed in the mid-1990s, M-theory posits that these seemingly distinct theories are actually different manifestations of a single, more fundamental theory. Here's an overview of M-theory.

Basics of M-Theory

Unification of Superstring Theories:

- **Five Superstring Theories:** Before M-theory, there were five consistent superstring theories: Type I, Type IIA, Type IIB, heterotic SO(32), and heterotic E8×E8. Each theory appeared different, with unique properties and structures.
- **Single Framework:** M-theory suggests that these five theories are just different aspects of the same underlying theory, connected through various dualities.

Eleven Dimensions:

- **Extra Dimension:** While superstring theories require ten dimensions, M-theory proposes an additional eleventh dimension. This extra dimension is critical for unifying the different string theories.
- **Compactification:** The extra dimensions are compactified into complex shapes, typically at incredibly small scales, making them unobservable at low energies.

Mathematical Structure

Membranes and Higher-Dimensional Objects:

- **Extended Objects:** In addition to one-dimensional strings, M-theory includes higher-dimensional objects known as membranes or branes. These can be two-dimensional (membranes) or even higher-dimensional (p-branes), where "p" indicates the number of spatial dimensions.
- **Brane Dynamics:** The inclusion of branes allows for a richer and more flexible theoretical structure, capable of describing a wider variety of physical phenomena.

Dualities:

- **String Dualities:** M-theory relies on various dualities that relate different superstring theories. For example, T-duality relates large and small compact dimensions, while S-duality connects strong and weak coupling constants.
- **U-Duality:** A more comprehensive duality in M-theory, known as U-duality, combines both T-duality and S-duality, further supporting the idea that all string theories are interconnected.

Physical Implications

Quantum Gravity:

- **Graviton:** M-theory naturally incorporates the graviton, the hypothetical quantum particle that mediates gravity. This makes it a strong candidate for a theory of quantum gravity, capable of describing gravitational interactions at quantum scales.
- **Black Holes and Singularities:** M-theory provides new insights into the nature of black holes and singularities, suggesting ways to resolve the infinities predicted by general relativity.

Cosmology:

- **Early Universe:** M-theory has implications for the early universe, including mechanisms for cosmic inflation and the formation of large-scale structures. It offers potential explanations for the Big Bang and the evolution of the cosmos.

Experimental Prospects

Indirect Evidence:

- **High-Energy Physics:** Direct experimental evidence for M-theory is challenging due to the small scale of the extra dimensions and the high energies required. However, researchers seek indirect evidence through high-energy physics experiments, such as those conducted at the Large Hadron Collider (LHC).
- **Cosmological Observations:** Observations of the cosmic microwave background, gravitational waves, and other cosmological phenomena may provide clues supporting the predictions of M-theory.

Challenges and Future Directions

Mathematical Complexity:

- **Advanced Techniques:** Understanding and developing M-theory requires advanced mathematical tools from algebraic geometry, topology, and quantum field theory. Ongoing research aims to refine these techniques and explore the full implications of the theory.
- **Unresolved Questions:** Many aspects of M-theory remain speculative, and significant theoretical and experimental work is needed to confirm its validity and uncover its full potential.

Duality Relationships

In string theory and M-theory, duality relationships are important in unifying different theories and revealing deep connections between seemingly distinct physical frameworks. These dualities show that different theories are actually different manifestations of a single, underlying theory. Here's an overview of key duality relationships in string theory and M-theory.

T-Duality

Basic Concept:

- **Large and Small Radius Equivalence:** T-duality is a symmetry that relates string theories compactified on circles of radius R to theories compactified on circles of radius $1/R$. This means that a string theory on a large spatial dimension is equivalent to a string theory on a small spatial dimension.
- **Implications for Strings:** In T-duality, the momentum modes of strings in one theory correspond to winding modes in the dual theory. This duality illustrates the concept that strings can wrap around compact dimensions.

Example:

- **Type IIA and Type IIB:** T-duality can relate Type IIA string theory compactified on a circle to Type IIB string theory compactified on a circle of the inverse radius, and vice versa.

S-Duality

Basic Concept:

- **Strong-Weak Coupling Duality:** S-duality is a relationship between theories with strong coupling constants and those with weak coupling constants. It suggests that a strongly interacting theory can be equivalent to a weakly interacting theory.
- **Implications for Gauge Theories:** S-duality often relates different types of gauge theories and can provide insights into non-perturbative aspects of these theories.

Example:

- **Type IIB String Theory:** S-duality relates Type IIB string theory at strong coupling to Type IIB string theory at weak coupling, implying that the theory is self-dual under S-duality.

U-Duality

Comprehensive Duality:

- **Combining T and S-Dualities:** U-duality is a more extensive duality that encompasses both T-duality and S-duality. It combines these symmetries to relate different superstring theories and M-theory.
- **Implications for M-Theory:** U-duality is important in the context of M-theory, revealing that various lower-dimensional theories are interconnected through a higher-dimensional framework.

Mirror Symmetry

Basic Concept:

- **Calabi-Yau Manifolds:** Mirror symmetry is a duality between pairs of Calabi-Yau manifolds. These manifolds are used in string theory to compactify extra dimensions. Mirror symmetry suggests that the physics of a string theory compactified on one Calabi-Yau manifold is equivalent to the physics of a string theory compactified on the mirror manifold.
- **Implications for Geometry and Physics:** Mirror symmetry has profound implications for both mathematics and physics, particularly in the study of

complex geometry and algebraic geometry. It provides tools for counting holomorphic curves and understanding the structure of Calabi-Yau spaces.

Example:

- **Type IIA and Type IIB:** Mirror symmetry can relate Type IIA string theory compactified on a Calabi-Yau manifold to Type IIB string theory compactified on the mirror Calabi-Yau manifold.

Electric-Magnetic Duality

Basic Concept:

- **Charge and Flux:** Electric-magnetic duality, or Montonen-Olive duality, is a type of S-duality in certain gauge theories. It relates the electric charges in one theory to magnetic monopoles in another, indicating that these theories are equivalent under the interchange of electric and magnetic fields.
- **Implications for Field Theories:** This duality provides insights into the non-perturbative structure of gauge theories and has implications for understanding confinement and duality in quantum field theory.

Examples and Implications

Connecting Different Theories:

- **Type I and Heterotic SO(32):** S-duality relates Type I string theory to the heterotic SO(32) string theory. This connection shows that these seemingly different theories describe the same physics under different coupling regimes.
- **Heterotic E8×E8 and M-Theory:** T-duality and U-duality link the heterotic E8×E8 string theory to M-theory compactified on certain manifolds, demonstrating the deep connections between these theories.

Experimental and Theoretical Challenges

Testing Dualities:

- **Indirect Evidence:** While direct experimental evidence for these dualities is challenging to obtain, researchers look for indirect signs through high-energy experiments and cosmological observations.
- **Mathematical Rigor:** Much of the evidence for dualities comes from mathematical consistency and theoretical insights. Advancing mathematical techniques and developing new tools are essential for further understanding these dualities.

Future Directions:

- **Quantum Gravity:** Dualities provide a framework for understanding quantum gravity and the unification of forces. They are crucial for exploring the non-perturbative aspects of string theory and M-theory.
- **Interdisciplinary Impact:** Insights from dualities influence various fields, including condensed matter physics, quantum field theory, and pure mathematics, highlighting their broad relevance.

Duality relationships in string theory and M-theory reveal connections between different theories, suggesting they're various manifestations of a single underlying framework.

CHAPTER 9: SUPERSYMMETRY

Basics of Supersymmetry

Supersymmetry (SUSY) is a theoretical framework in particle physics that posits a symmetry between bosons (particles that carry forces) and fermions (particles that make up matter). It extends the Standard Model of particle physics and addresses several unresolved issues.

Core Concept
Symmetry Between Particles:

- **Bosons and Fermions:** Supersymmetry proposes that each boson has a corresponding fermion superpartner and vice versa. For example, the electron (a fermion) would have a superpartner called the selectron (a boson), and the photon (a boson) would have a superpartner called the photino (a fermion).

Mathematical Structure:

- **Supercharges:** The mathematical foundation of supersymmetry involves operators known as supercharges. These operators transform bosons into fermions and fermions into bosons, thus connecting particles with their superpartners.
- **Supermultiplets:** Particles and their superpartners are grouped into supermultiplets. Each supermultiplet contains both bosons and fermions, ensuring the symmetry is maintained.

Motivations for Supersymmetry
Hierarchy Problem:

- **Stabilizing the Higgs Boson:** One of the major motivations for SUSY is the hierarchy problem. In the Standard Model, quantum corrections can make the mass of the Higgs boson extremely large. Supersymmetry helps cancel out these corrections, stabilizing the Higgs mass at a manageable level.

Unification of Forces:

- **Gauge Coupling Unification:** Supersymmetry allows for the unification of the three gauge forces (electromagnetic, weak, and strong) at high energy scales. In SUSY models, the running of the gauge couplings naturally converges at a single point, supporting the idea of a grand unified theory (GUT).

Dark Matter:

- **Lightest Supersymmetric Particle:** Supersymmetry provides a candidate for dark matter in the form of the lightest supersymmetric particle (LSP). The LSP is stable and weakly interacting, making it a suitable dark matter candidate.

Supersymmetric Models

Minimal Supersymmetric Standard Model (MSSM):

- **Extending the Standard Model:** The MSSM is the simplest extension of the Standard Model that incorporates supersymmetry. It introduces superpartners for all Standard Model particles and includes two Higgs doublets instead of one, leading to additional Higgs bosons.
- **Soft SUSY Breaking:** To reconcile SUSY with the observed lack of superpartners at low energies, the MSSM includes mechanisms for soft SUSY breaking. These mechanisms break supersymmetry at low energy scales without introducing large quantum corrections.

Experimental Search for Supersymmetry

Collider Experiments:

- **Large Hadron Collider (LHC):** The LHC is at the forefront of the search for supersymmetry. Physicists look for signals of superpartners, such as missing energy signatures that could indicate the presence of the LSP escaping detection.
- **Mass Constraints:** So far, no superpartners have been definitively detected, leading to constraints on their masses. This has pushed the potential discovery scale for SUSY particles to higher energies.

Indirect Evidence:

- **Precision Measurements:** Supersymmetry can affect various precision measurements, such as the anomalous magnetic moment of the muon and rare decay processes. Deviations from Standard Model predictions in these measurements could hint at SUSY.

Theoretical and Practical Implications

Advancing Theoretical Physics:

- **String Theory and Quantum Gravity:** Supersymmetry is a crucial component of string theory, helping to ensure the theory's mathematical consistency. It also plays a role in potential theories of quantum gravity, aiming to unify all fundamental forces.
- **Mathematical Elegance:** SUSY introduces new symmetries and structures in theoretical physics, offering elegant solutions to some of the most pressing problems in the field.

Future Prospects:

- **Next-Generation Colliders:** Future high-energy colliders could explore energy scales beyond the reach of the LHC, potentially discovering superpartners and providing direct evidence for SUSY.
- **Refining Models:** Continued theoretical work aims to refine supersymmetric models, incorporating the latest experimental data and improving predictions for future experiments.

Supersymmetry is a powerful theoretical framework that extends the Standard Model by introducing a symmetry between bosons and fermions. It addresses significant issues like the hierarchy problem, gauge coupling unification, and dark matter. Ongoing experimental searches and theoretical advancements continue to explore the validity and implications of this elegant theory.

Supersymmetry in String Theory

Supersymmetry is integral to string theory, enhancing its mathematical consistency and physical relevance. In string theory, SUSY provides a symmetry between bosons (force-carrying particles) and fermions (matter particles), offering several advantages and resolving key issues.

Core Concepts

Bosons and Fermions:

- **Superpartners:** In supersymmetric string theory, each particle has a superpartner: bosons have fermionic superpartners and fermions have bosonic ones. For example, the electron has a superpartner called the selectron, and the photon has a superpartner called the photino.

Supercharges and Supermultiplets:

- **Supercharges:** Supersymmetry is mathematically described by supercharges, which transform bosons into fermions and vice versa.

- **Supermultiplets:** Particles and their superpartners are grouped into supermultiplets, maintaining the symmetry.

Benefits in String Theory

Mathematical Consistency:

- **Anomaly Cancellation:** Supersymmetry helps cancel anomalies, ensuring the theory's consistency. This is crucial for avoiding mathematical inconsistencies that can arise in quantum field theories.
- **Stability:** SUSY helps stabilize the theory by preventing the runaway behavior of quantum corrections, which can otherwise lead to unrealistic predictions.

Unification of Forces:

- **Gauge Coupling Unification:** SUSY facilitates the unification of fundamental forces at high energy scales, supporting the concept of a grand unified theory (GUT).

Quantum Gravity:

- **Incorporation in M-Theory:** Supersymmetry is a key feature of M-theory, the proposed overarching theory that unifies all string theories. M-theory requires eleven dimensions and includes various types of branes in addition to strings.

Experimental Search

Collider Experiments:

- **LHC Searches:** The Large Hadron Collider (LHC) is actively searching for supersymmetric particles. Although none have been definitively detected, ongoing experiments aim to find evidence of these superpartners.

Indirect Evidence:

- **Precision Measurements:** Supersymmetry may manifest in precision measurements, such as the anomalous magnetic moment of the muon, offering indirect hints of its existence.

Superpartners

In supersymmetry, every particle in the Standard Model has a corresponding partner called a superpartner. These superpartners differ in spin by half a unit and help solve various theoretical problems, providing deeper insights into the nature of the universe.

Basics of Superpartners

Bosons and Fermions:

- **Bosons:** Particles that carry forces, such as photons and gluons. They have integer spins (0, 1, 2, etc.).
- **Fermions:** Particles that make up matter, such as electrons and quarks. They have half-integer spins (1/2, 3/2, etc.).

Superpartners:

- **Fermionic Superpartners (Sfermions):** Each fermion has a bosonic superpartner. For instance, the electron (a fermion) has a superpartner called the selectron (a boson).
- **Bosonic Superpartners (Gauginos and Higgsinos):** Each boson has a fermionic superpartner. For example, the photon (a boson) has a superpartner called the photino (a fermion).

Specific Examples

Leptons and Sleptons:

- **Electron (e-) and Selectron ($e \sim \tilde{e} e \sim$):** The electron, a negatively charged lepton, has a scalar superpartner called the selectron, which has no spin.
- **Muon (μ) and Smuon ($\mu \sim \tilde{\mu} \mu \sim$):** The muon, a heavier cousin of the electron, has a superpartner called the smuon.

Quarks and Squarks:

- **Up Quark (u) and Sup ($u \sim \tilde{u} u \sim$):** The up quark, a fundamental constituent of protons, has a superpartner called the sup.
- **Down Quark (d) and Sdown ($d \sim \tilde{d} d \sim$):** The down quark, another fundamental constituent of protons and neutrons, has a superpartner called the sdown.

Gauge Bosons and Gauginos:

- **Photon (γ) and Photino (γ~\tilde{γ}γ~):** The photon, which mediates electromagnetic interactions, has a fermionic superpartner called the photino.
- **Gluon (g) and Gluino (g~\tilde{g}g~):** The gluon, which mediates strong interactions, has a superpartner called the gluino.

Higgs Bosons and Higgsinos:

- **Higgs Boson (H) and Higgsino (H~\tilde{H}H~):** The Higgs boson, responsible for giving mass to other particles, has a fermionic superpartner called the Higgsino.

Importance of Superpartners
Solving the Hierarchy Problem:

- **Stabilizing the Higgs Mass:** Quantum corrections can make the Higgs boson's mass extremely large. Superpartners help cancel out these corrections, stabilizing the Higgs mass at a manageable scale.

Gauge Coupling Unification:

- **Unification at High Energies:** Superpartners allow for the unification of the electromagnetic, weak, and strong forces at high energy scales. This supports the idea of a grand unified theory (GUT).

Dark Matter Candidate:

- **Lightest Supersymmetric Particle (LSP):** In many SUSY models, the lightest supersymmetric particle is stable and weakly interacting, making it a strong candidate for dark matter.

Experimental Searches
Collider Experiments:

- **Large Hadron Collider (LHC):** The LHC is searching for evidence of superpartners. If superpartners exist, they should be produced in high-energy collisions, potentially leaving detectable signatures.

Indirect Evidence:

- **Precision Measurements:** Deviations from Standard Model predictions in precision measurements, such as the magnetic moment of the muon, could indicate the influence of superpartners.

Current Status and Future Prospects

No Direct Detection Yet:

- **Mass Constraints:** Despite extensive searches, superpartners have not yet been directly detected, leading to constraints on their masses. Researchers continue to refine their models and search for superpartners at higher energy scales.

Theoretical Developments:

- **Refining SUSY Models:** Ongoing theoretical work aims to refine supersymmetric models, incorporating the latest experimental data and improving predictions for future experiments.

Experimental Challenges

Supersymmetry is a compelling theoretical framework that proposes a symmetry between bosons and fermions, offering solutions to several outstanding issues in particle physics. However, despite its theoretical appeal, detecting superpartners and confirming SUSY experimentally presents significant challenges. Here are the primary experimental challenges faced in the search for supersymmetry:

High Energy Requirements

Particle Accelerators:

- **Energy Thresholds:** Supersymmetric particles, if they exist, are expected to be much heavier than the known particles of the Standard Model. Detecting them requires particle collisions at extremely high energies, often beyond the current capabilities of existing accelerators.
- **Large Hadron Collider (LHC):** The LHC, the most powerful particle accelerator to date, has been searching for SUSY particles. While it has set constraints on the masses of these particles, it has yet to find direct evidence for them.

Signal Detection and Background Noise

Complex Signatures:

- **Similar to Standard Model Events:** The decay patterns and signatures of SUSY particles can be very similar to those of Standard Model particles, making it difficult to distinguish between the two.
- **Background Noise:** High-energy collisions produce a vast amount of data, with numerous particles and interactions occurring simultaneously.

Isolating potential SUSY signals from this background noise is a significant challenge.

Missing Energy:

- **Lightest Supersymmetric Particle (LSP):** Many SUSY models predict the existence of a stable, weakly interacting lightest supersymmetric particle, which would escape detection, similar to neutrinos. This would manifest as missing transverse energy in the detector, but such signals can also be produced by other processes, complicating the identification of SUSY events.

Theoretical Uncertainty

Parameter Space:

- **Vast Landscape:** SUSY models contain many free parameters, leading to a vast theoretical landscape. Determining the most promising regions of parameter space to search for SUSY signals requires careful consideration and extensive computational modeling.
- **Model Variations:** Different SUSY models (e.g., Minimal Supersymmetric Standard Model, Next-to-Minimal Supersymmetric Standard Model) predict different particle masses and interactions, adding to the complexity of the search.

Current Experimental Constraints

Mass Limits:

- **Exclusion Regions:** The LHC has ruled out certain regions of the SUSY parameter space, setting lower mass limits for various superpartners. For example, gluinos (the superpartners of gluons) are excluded below certain mass thresholds.
- **Refining Searches:** These constraints help refine theoretical models and focus future searches, but they also highlight the need for even higher energy experiments.

Future Prospects

Next-Generation Colliders:

- **Higher Energies:** Future particle colliders, such as the proposed Future Circular Collider (FCC) or the International Linear Collider (ILC), aim to achieve higher collision energies and luminosities, increasing the chances of detecting SUSY particles.

- **Precision Measurements:** These colliders will also provide more precise measurements of known particles, potentially revealing indirect effects of supersymmetry.

Indirect Evidence:

- **Cosmology and Astrophysics:** Observations in cosmology and astrophysics, such as dark matter searches and precision measurements of cosmic microwave background radiation, may provide indirect evidence for SUSY. The properties of dark matter candidates predicted by SUSY could match astrophysical observations.
- **Anomalies in Precision Measurements:** Deviations from Standard Model predictions in precision measurements, such as the magnetic moment of the muon, could indicate the presence of superpartners.

So while supersymmetry offers an elegant extension to the Standard Model with solutions to significant theoretical problems, experimental verification remains challenging. High energy requirements, complex signatures, vast parameter spaces, and the need for next-generation colliders are key hurdles. Nonetheless, ongoing research and technological advancements continue to push the boundaries, keeping the search for supersymmetry a central focus in particle physics.

CHAPTER 10: STRING THEORY IN MODERN PHYSICS

Impact on Cosmology

String theory impacts modern cosmology by providing new frameworks and explanations for the universe's most profound mysteries. Here's how string theory influences our understanding of cosmology.

The Early Universe

Inflation:

- **Cosmic Inflation:** String theory offers potential mechanisms for cosmic inflation, the rapid expansion of the universe moments after the Big Bang. This theory helps explain the uniformity and flatness of the observable universe.
- **Brane Inflation:** One such mechanism involves brane inflation, where the separation and collision of branes (multidimensional objects in string theory) drive the inflationary expansion. This idea extends traditional inflationary models by incorporating higher dimensions.

Structure Formation

Cosmic Strings:

- **Topological Defects:** String theory predicts the existence of cosmic strings, one-dimensional topological defects formed during phase transitions in the early universe. These cosmic strings could have played a role in the formation of large-scale structures, such as galaxies and clusters.
- **Observational Signatures:** If cosmic strings exist, they would leave distinct imprints on the cosmic microwave background (CMB) and the distribution of galaxies. Detecting these signatures would provide evidence for string theory's role in cosmology.

Dark Matter and Dark Energy

Supersymmetry:

- **Dark Matter Candidates:** String theory often incorporates supersymmetry, which predicts the existence of stable, weakly interacting particles suitable as dark matter candidates. These particles, such as the neutralino, could make up the dark matter halo around galaxies.
- **Dark Energy:** String theory also provides frameworks for understanding dark energy, the mysterious force driving the accelerated expansion of the

universe. The theory's extra dimensions and vacuum states could influence the cosmological constant and the dynamics of dark energy.

Multiverse and Extra Dimensions

Multiverse Hypothesis:

- **String Landscape:** String theory's vast number of possible vacuum states, known as the string landscape, suggests the existence of a multiverse. Each vacuum state could correspond to a different universe with its own physical laws.
- **Anthropic Principle:** The multiverse hypothesis supports the anthropic principle, which posits that we observe certain physical constants because they allow for life. In a multiverse, regions with different constants exist, but we inhabit one of the few regions where conditions are right for life.

Extra Dimensions:

- **Compactification:** String theory requires extra dimensions beyond the familiar four (three spatial and one temporal). These extra dimensions are compactified into tiny shapes, influencing the physical properties of our universe.
- **Brane World Scenarios:** In some models, our observable universe is a 3-dimensional brane embedded in a higher-dimensional space. This idea can explain various cosmological phenomena, such as the hierarchy problem and the nature of gravity.

Black Holes and Singularities

Black Hole Physics:

- **Hawking Radiation:** String theory provides insights into black hole physics, including the nature of Hawking radiation and the information paradox. The theory suggests that information is not lost in black holes but encoded in subtle ways.
- **Microstates:** In string theory, black holes can be described by a vast number of microstates, offering a statistical explanation for black hole entropy. This approach helps resolve singularities and provides a quantum description of black holes.

Future Prospects

Cosmological Observations:

- **CMB and Large-Scale Structure:** Ongoing and future observations of the cosmic microwave background and large-scale structure of the universe will test string theory's predictions. These observations may provide indirect evidence for extra dimensions, cosmic strings, and other string-theoretic phenomena.

- **Gravitational Waves:** The detection of gravitational waves from cosmic events, such as black hole mergers, offers new opportunities to test string theory. These waves could carry signatures of extra dimensions or other string-theoretic effects.

Theoretical Developments:

- **Advancing Models:** Researchers continue to refine string-theoretic models to better match cosmological observations. This includes developing more precise calculations and exploring new aspects of string theory that could impact cosmology.

String theory impacts cosmology by offering new explanations for the early universe, structure formation, dark matter, dark energy, and the nature of black holes. Its frameworks for extra dimensions and the multiverse open up new avenues for understanding the cosmos. Ongoing observations and theoretical developments will continue to test and refine these ideas.

Black Holes and Strings

Black holes and string theory intersect in fascinating ways, offering insights into the nature of gravity, spacetime, and quantum mechanics. Here's an exploration of how string theory impacts our understanding of black holes.

Microstates and Black Hole Entropy

Black Hole Entropy:

- **Bekenstein-Hawking Entropy:** In classical general relativity, black holes have entropy proportional to the area of their event horizons. This is known as the Bekenstein-Hawking entropy, given by $S = kA / (4Lp^2)$ where A is the area, k is Boltzmann's constant, and Lp is the Planck length.
- **String Theory Microstates:** String theory explains this entropy in terms of microstates. A black hole's entropy counts the number of distinct quantum states (microstates) that correspond to the same macroscopic black hole. These microstates arise from different configurations of strings and branes, which form the black hole.

Hawking Radiation and Information Paradox

Hawking Radiation:

- **Quantum Emission:** Stephen Hawking showed that black holes emit radiation due to quantum effects near the event horizon, leading to black hole evaporation over time. This radiation is known as Hawking radiation.
- **Information Paradox:** Hawking radiation implies that information about the matter that formed the black hole could be lost as the black hole evaporates, violating the principle of quantum mechanics that information cannot be destroyed.

String Theory's Resolution:

- **Information Preservation:** String theory suggests that information is not lost but rather encoded in the radiation or in subtle correlations between emitted particles. Various string-theoretic models, such as the holographic principle, indicate that information is preserved.

Holographic Principle and AdS/CFT Correspondence

Holographic Principle:

- **Information on Boundaries:** The holographic principle posits that all information contained within a volume of space can be described by a theory on the boundary of that space. For black holes, this means the information about the interior can be encoded on the event horizon.

AdS/CFT Correspondence:

- **Duality:** Proposed by Juan Maldacena, the AdS/CFT correspondence is a duality between a string theory formulated in an anti-de Sitter (AdS) space and a conformal field theory (CFT) on its boundary. This duality provides a powerful tool for studying black holes.
- **Black Hole Solutions:** In the AdS/CFT framework, a black hole in AdS space can be described by a CFT, allowing physicists to study black hole thermodynamics and quantum properties using well-understood field theory techniques.

Black Hole Microstates and String Theory

D-Branes:

- **Building Blocks:** In string theory, D-branes are fundamental objects on which strings can end. Black holes can be modeled as configurations of intersecting D-branes.
- **Microstate Counting:** By counting the possible configurations of these D-branes, string theorists can account for the entropy of black holes, matching the Bekenstein-Hawking formula.

Fuzzball Proposal:

- **No Singularities:** The fuzzball proposal suggests that black holes are composed of a vast number of microstates, each corresponding to a different stringy configuration. Instead of a singularity, the interior of a black hole is a complex, stringy structure with no singular point.
- **Surface Structure:** According to this proposal, the traditional event horizon is replaced by a "fuzzy" surface, resolving singularities and information loss issues.

Observational Implications

Gravitational Waves:

- **Black Hole Mergers:** Observations of gravitational waves from black hole mergers provide a unique opportunity to test predictions from string theory. The properties of the waves, such as their echoes, might reveal hints of stringy effects or deviations from classical general relativity.

Event Horizon Telescope:

- **Imaging Black Holes:** The Event Horizon Telescope (EHT) aims to image the event horizons of black holes. Detailed observations could provide data to test string-theoretic models, such as the fuzzball proposal or other deviations from the classical black hole picture.

Future Directions

Theoretical Advances:

- **Refining Models:** Researchers continue to develop more precise string-theoretic models of black holes, improving our understanding of microstates, entropy, and quantum properties.
- **Quantum Gravity:** Insights from string theory contribute to the broader quest for a theory of quantum gravity, aiming to unify general relativity and quantum mechanics.

Experimental Searches:

- **High-Energy Collisions:** Future particle collider experiments might produce microscopic black holes, providing a direct way to test string theory predictions in a controlled laboratory setting.
- **Cosmological Observations:** Advances in observational cosmology, such as more sensitive gravitational wave detectors and next-generation telescopes, will provide further data to test and refine string-theoretic predictions about black holes.

In short, string theory impacts our understanding of black holes by offering explanations for black hole entropy, resolving the information paradox, and providing new frameworks through the holographic principle and AdS/CFT correspondence. Ongoing theoretical and observational efforts continue to test and refine these ideas, improving our understanding of the universe's most enigmatic objects.

String Theory in Particle Physics

String theory has made significant contributions to particle physics by providing a more comprehensive framework for understanding the fundamental particles and forces of nature. Here's a look at how string theory impacts particle physics.

Fundamental Concepts
Strings as Building Blocks:

- **One-Dimensional Strings:** In string theory, the fundamental entities are not point-like particles but one-dimensional strings. These strings can vibrate at different frequencies, and each vibrational mode corresponds to a different particle.
- **Types of Strings:** Strings can be open (with two endpoints) or closed (forming loops). Each type has unique properties and implications for particle physics.

Unification of Forces
Incorporating Gravity:

- **Graviton:** One of the vibrational modes of a closed string corresponds to the graviton, the hypothetical quantum particle that mediates the gravitational force. This naturally incorporates gravity into the quantum framework, addressing a significant gap in the Standard Model.
- **Quantum Gravity:** By including gravity, string theory aims to unify all fundamental forces—gravitational, electromagnetic, weak, and strong interactions—within a single theoretical framework.

Gauge Bosons:

- **Force Carriers:** Other vibrational modes of strings correspond to gauge bosons, the particles that mediate the fundamental forces. For example, photons, W and Z bosons, and gluons are represented by specific string vibrations.

- **Gauge Symmetry:** String theory supports gauge symmetries, which are fundamental to the Standard Model of particle physics. These symmetries dictate the interactions between particles and the forces they experience.

Supersymmetry (SUSY)

Superpartners:

- **Bosons and Fermions:** Supersymmetry posits that each particle has a corresponding superpartner with a spin differing by half a unit. For example, fermions like electrons have bosonic superpartners called selectrons, and bosons like photons have fermionic superpartners called photinos.
- **Solving Theoretical Problems:** SUSY helps address several issues in particle physics, such as stabilizing the Higgs boson's mass and enabling the unification of coupling constants at high energy scales.

Minimal Supersymmetric Standard Model (MSSM):

- **Extending the Standard Model:** The MSSM is the simplest supersymmetric extension of the Standard Model. It predicts new particles, including superpartners, that could be detected in high-energy experiments.

Extra Dimensions

Compactification:

- **Additional Dimensions:** String theory requires extra spatial dimensions beyond the familiar three. Typically, these extra dimensions are compactified into tiny shapes, such as Calabi-Yau manifolds, making them unobservable at low energies.
- **Impact on Particle Properties:** The shape and size of these compactified dimensions influence the properties of particles, including their masses and interaction strengths.

Brane World Scenarios:

- **Higher-Dimensional Branes:** In some models, our universe is a 3-dimensional brane embedded in a higher-dimensional space. This framework can explain various phenomena, such as the relative weakness of gravity compared to other forces.

Experimental Searches and Evidence

Collider Experiments:

- **Large Hadron Collider (LHC):** The LHC is at the forefront of searching for evidence of string theory. Researchers look for superpartners, extra dimensions, and deviations from the Standard Model predictions.
- **Signatures of Supersymmetry:** While no superpartners have been definitively detected, ongoing experiments aim to discover these particles, which would provide strong evidence for SUSY and string theory.

Indirect Evidence:

- **Precision Measurements:** Precision measurements, such as the anomalous magnetic moment of the muon and rare particle decays, could reveal indirect effects of string theory.
- **Cosmological Observations:** Observations of the cosmic microwave background, gravitational waves, and large-scale structure of the universe may provide clues supporting the existence of extra dimensions or other string-theoretic effects.

Theoretical Advances
Dualities:

- **String Dualities:** String theory includes various dualities, such as T-duality and S-duality, that relate different string theories and reveal deep connections between seemingly distinct models. These dualities help unify different approaches and provide a more cohesive theoretical framework.
- **AdS/CFT Correspondence:** The AdS/CFT correspondence is a powerful duality that connects a string theory in a higher-dimensional anti-de Sitter (AdS) space with a conformal field theory (CFT) on its boundary. This duality has profound implications for understanding quantum gravity and strongly interacting systems.

String Phenomenology:

- **Model Building:** String phenomenology involves constructing realistic models based on string theory that match observed particle physics phenomena. This includes developing compactification schemes and exploring different configurations of branes and strings.

Challenges and Future Directions
Testing Predictions:

- **High Energy Requirements:** Detecting direct evidence for string theory often requires extremely high energy scales, which may be beyond the reach of current experiments. Future colliders and advanced observational techniques are necessary to probe these predictions.

- **Parameter Space:** String theory's vast parameter space presents challenges for making specific predictions. Researchers continue to refine models and focus on the most promising regions.

Mathematical Complexity:

- **Advanced Mathematics:** String theory relies on complex mathematical concepts from algebraic geometry, topology, and quantum field theory. Ongoing advancements in these areas are crucial for further developing and testing the theory.

Interdisciplinary Impact:

- **Broad Applications:** Insights from string theory influence various fields, including condensed matter physics, cosmology, and pure mathematics. This interdisciplinary impact highlights the theory's potential to revolutionize our understanding of fundamental physics.

String Theory vs. Standard Model

While the Standard Model has been extremely successful in explaining a wide range of phenomena, string theory aims to offer a more comprehensive and unified approach that addresses some of the limitations of the Standard Model. Here's a comparison of string theory and the Standard Model.

The Standard Model
Core Principles:

- **Particles and Forces:** The Standard Model describes three of the four fundamental forces: electromagnetic, weak, and strong interactions. It does not include gravity.
- **Elementary Particles:** The model categorizes elementary particles into quarks, leptons, and gauge bosons. Quarks and leptons make up matter, while gauge bosons mediate the forces.
  - **Quarks:** Up, down, charm, strange, top, and bottom.
  - **Leptons:** Electron, muon, tau, and their corresponding neutrinos.
  - **Gauge Bosons:** Photon (electromagnetic force), W and Z bosons (weak force), and gluons (strong force).
- **Higgs Mechanism:** The Higgs boson, discovered in 2012, provides particles with mass through the Higgs mechanism.

Mathematical Framework:

- **Quantum Field Theory (QFT):** The Standard Model is formulated as a quantum field theory, combining special relativity and quantum mechanics.
- **Gauge Symmetry:** The model is based on the gauge symmetries SU(3)×SU(2)×U(1), corresponding to the strong, weak, and electromagnetic interactions, respectively.

Successes:

- **Predictive Power:** The Standard Model has accurately predicted numerous particle interactions and properties, confirmed by experimental results.
- **Experimental Validation:** Discoveries such as the W and Z bosons, top quark, and Higgs boson have validated the model's predictions.

Limitations of the Standard Model

Incompleteness:

- **Gravity:** The Standard Model does not include gravity, which is described separately by general relativity.
- **Dark Matter and Dark Energy:** The model does not account for dark matter and dark energy, which make up most of the universe's mass-energy content.
- **Neutrino Masses:** Initially, the Standard Model predicted massless neutrinos, but experiments have shown that neutrinos have tiny masses.

Hierarchies and Fine-Tuning:

- **Hierarchy Problem:** The large difference between the weak force and gravity scales (the Planck scale) requires fine-tuning, leading to the hierarchy problem.
- **Strong CP Problem:** The model does not naturally explain why the strong force does not break charge-parity (CP) symmetry significantly.

String Theory

Core Principles:

- **Strings as Fundamental Entities:** String theory posits that the fundamental constituents of the universe are one-dimensional strings rather than point particles. These strings can vibrate at different frequencies, and each mode of vibration corresponds to a different particle.
- **Incorporation of Gravity:** One of the vibrational modes of the string corresponds to the graviton, a hypothetical quantum particle that mediates gravity, naturally incorporating gravity into the framework.

Mathematical Framework:

- **Extra Dimensions:** String theory requires additional spatial dimensions beyond the familiar three. Typically, it proposes ten dimensions in superstring theory or eleven in M-theory, with the extra dimensions compactified into tiny shapes.
- **Supersymmetry:** String theory often incorporates supersymmetry, which posits that every particle has a corresponding superpartner, helping to resolve various theoretical issues such as the hierarchy problem.

Advantages:

- **Unification of Forces:** String theory aims to unify all fundamental forces, including gravity, into a single theoretical framework.
- **Quantum Gravity:** It provides a consistent theory of quantum gravity, addressing the limitations of both the Standard Model and general relativity.

Theoretical Elegance:

- **Gauge Symmetries and Dualities:** String theory exhibits rich mathematical structures, including various dualities that relate different string theories and reveal deep connections between seemingly distinct models.

Challenges and Criticisms

Experimental Verification:

- **High Energy Requirements:** Direct experimental evidence for string theory often requires energy scales much higher than current or foreseeable particle accelerators can achieve.
- **Indirect Evidence:** Researchers look for indirect evidence through phenomena such as cosmic strings, deviations in particle physics experiments, and cosmological observations, but none have definitively confirmed string theory yet.

Mathematical Complexity:

- **Parameter Space:** The vast number of possible solutions (vacua) in string theory, known as the "landscape problem," makes it challenging to make specific, testable predictions.
- **Advanced Mathematics:** Understanding and developing string theory requires sophisticated mathematical tools from algebraic geometry, topology, and quantum field theory.

Future Prospects

Experimental Advances:

- **Next-Generation Colliders:** Future high-energy colliders, such as the proposed Future Circular Collider (FCC) or the International Linear Collider (ILC), may probe energy scales where evidence for supersymmetry or extra dimensions could be found.
- **Cosmological Observations:** Advances in observational cosmology, such as more sensitive gravitational wave detectors and next-generation telescopes, may provide further data to test string-theoretic predictions.

Theoretical Developments:

- **Refining Models:** Researchers continue to refine string-theoretic models to better match observed phenomena in particle physics and cosmology.
- **Interdisciplinary Impact:** Insights from string theory contribute to various fields, including condensed matter physics, quantum information theory, and pure mathematics, highlighting its broad relevance.

AdS/CFT Correspondence

The AdS/CFT correspondence, proposed by Juan Maldacena in 1997, is a concept in theoretical physics that connects two seemingly different theories: a type of string theory formulated in anti-de Sitter (AdS) space and a conformal field theory (CFT) defined on the boundary of that space. This duality provides insights into quantum gravity and strongly interacting quantum field theories. Below we have a look at the AdS/CFT correspondence.

Basic Concepts

Anti-de Sitter (AdS) Space:

- **Curved Spacetime:** AdS space is a mathematical model of a universe with a constant negative curvature, which contrasts with the flat or positively curved spacetimes more familiar from general relativity.
- **Higher Dimensions:** In the context of the AdS/CFT correspondence, AdS space typically has one more spatial dimension than the CFT. For example, AdS5 (five-dimensional AdS space) is related to a four-dimensional CFT.

Conformal Field Theory (CFT):

- **Scale Invariance:** A CFT is a quantum field theory that is invariant under conformal transformations, which include scalings, rotations, and translations.
- **Boundary Theory:** In the AdS/CFT correspondence, the CFT is defined on the boundary of the AdS space, which has one fewer dimension.

The Correspondence

Duality:

- **Equivalence:** The AdS/CFT correspondence posits that a string theory (or M-theory) formulated in the bulk of AdS space is equivalent to a CFT on the boundary of that space. This means that every physical quantity in the CFT can be mapped to a corresponding quantity in the AdS space and vice versa.
- **Holographic Principle:** This duality is a realization of the holographic principle, which suggests that the information contained within a volume of space can be represented by a theory on its boundary.

Implications for Quantum Gravity:

- **Quantum Gravity in AdS:** The correspondence provides a framework for studying quantum gravity in AdS space by using the well-understood tools of CFT.
- **Black Hole Information Paradox:** Insights from the AdS/CFT correspondence help address the black hole information paradox by suggesting that information is preserved and encoded on the boundary.

Applications and Impact

Strongly Coupled Systems:

- **Calculations in QCD:** The correspondence has applications in understanding strongly coupled systems, such as quark-gluon plasma in quantum chromodynamics (QCD). It provides computational techniques to study these systems non-perturbatively.
- **Condensed Matter Physics:** AdS/CFT has also been applied to problems in condensed matter physics, including high-temperature superconductivity and quantum phase transitions.

Mathematical Insights:

- **String Theory and Geometry:** The correspondence offers deep mathematical insights into the geometry of string theory and the structure of space-time.

- **Exact Solutions:** It allows for the calculation of exact results in certain quantum field theories that were previously intractable.

Future Prospects

Experimental Tests:

- **Cosmology and Black Holes:** While direct experimental verification is challenging, indirect evidence through cosmological observations and the study of black holes may provide support for the correspondence.
- **Advanced Theories:** Ongoing research aims to extend the AdS/CFT correspondence to more general spacetimes and other types of field theories.

The AdS/CFT correspondence is a theoretical concept that bridges string theory and quantum field theory, offering insights into quantum gravity and strongly coupled systems. Its implications continue to influence various areas of theoretical physics and mathematics.

CHAPTER 11: EXPERIMENTAL APPROACHES

Collider Experiments

Collider experiments are at the forefront of modern physics, providing important insights into the fundamental particles and forces that govern our universe. These experiments involve accelerating particles to extremely high energies and then colliding them, allowing scientists to study the resulting interactions and particles produced.

The Basics of Collider Experiments
Particle Accelerators:

- **High Energy Collisions:** Particle accelerators, such as the Large Hadron Collider (LHC) at CERN, accelerate protons or heavy ions to near light speeds. When these particles collide, they create conditions similar to those just after the Big Bang, enabling the study of fundamental physics.
- **Detectors:** Surrounding the collision points are large detectors like ATLAS, CMS, and ALICE. These detectors track the particles produced in the collisions, recording data about their properties, such as momentum, energy, and charge.

Goals of Collider Experiments:

- **Discover New Particles:** One of the primary goals is to discover new particles predicted by theoretical models, such as supersymmetric particles, extra dimensions, or even mini black holes.
- **Understand Fundamental Forces:** Colliders help scientists understand the fundamental forces, particularly the strong and weak nuclear forces, by studying how particles interact at high energies.
- **Test Theories:** Collider experiments test the predictions of the Standard Model and beyond, including string theory, which predicts various new phenomena that could manifest at high energy scales.

Key Achievements
Higgs Boson Discovery:

- **2012 Milestone:** The LHC's discovery of the Higgs boson in 2012 was a monumental achievement, confirming the last missing piece of the Standard Model. The Higgs boson gives other particles mass through the Higgs mechanism.

Testing Supersymmetry:

- **Search for Superpartners:** The LHC is searching for evidence of supersymmetry. SUSY predicts the existence of superpartners for each Standard Model particle, which could help solve the hierarchy problem and provide a candidate for dark matter.
- **Current Status:** Despite extensive searches, no superpartners have been conclusively detected yet, pushing the possible mass range for these particles higher.

Future Prospects

High Luminosity LHC:

- **Increasing Data:** The High Luminosity LHC (HL-LHC) aims to increase the LHC's luminosity, significantly boosting the number of collisions. This will improve the chances of detecting rare processes and new particles.
- **Precision Measurements:** The HL-LHC will also enable more precise measurements of known particles, providing further tests of the Standard Model and potential hints of new physics.

Next-Generation Colliders:

- **Future Circular Collider (FCC):** Proposals like the FCC aim to achieve even higher energies and luminosities. These next-generation colliders could probe deeper into the fabric of the universe, potentially uncovering evidence for string theory, extra dimensions, or other exotic phenomena.
- **International Linear Collider (ILC):** Another proposal, the ILC, would provide a cleaner environment for precision studies of the Higgs boson and other particles, complementing the discoveries of hadron colliders.

Challenges and Technological Innovations

Technical Hurdles:

- **High Energy Requirements:** Achieving the necessary collision energies and maintaining stable operation of particle accelerators present significant technical challenges.
- **Data Processing:** Collider experiments generate vast amounts of data, requiring advanced data processing and analysis techniques to identify and study interesting events.

Innovative Solutions:

- **Detector Upgrades:** Continuous upgrades to detectors and accelerators enhance their sensitivity and resolution, improving the quality and quantity of data collected.
- **Computational Advances:** Advances in computational power and algorithms enable more efficient data analysis, helping scientists sift through massive datasets to find significant signals.

Astrophysical Observations

Astrophysical observations are important for testing theories in modern physics, including string theory. By studying the universe at large scales, scientists gather data that can confirm or challenge theoretical predictions. Here's how these observations contribute to our understanding of the cosmos.

Cosmic Microwave Background (CMB)
Early Universe:

- **Big Bang Afterglow:** The CMB is the residual radiation from the Big Bang, providing a snapshot of the early universe.
- **Planck Satellite:** Detailed measurements from the Planck satellite offer insights into the universe's composition, age, and rate of expansion, testing models of cosmology and particle physics.

Gravitational Waves
Ripples in Spacetime:

- **LIGO and Virgo:** Detectors like LIGO and Virgo have observed gravitational waves from merging black holes and neutron stars. These observations confirm general relativity's predictions and provide new ways to probe the structure of spacetime.
- **Potential for String Theory:** Gravitational waves could reveal signatures of extra dimensions or cosmic strings, offering indirect evidence for string theory.

Dark Matter and Dark Energy
Cosmic Mysteries:

- **Dark Matter:** Observations of galaxy rotation curves and gravitational lensing indicate the presence of dark matter. Experiments like those using the Hubble Space Telescope and ground-based observatories aim to map dark matter distribution.

- **Dark Energy:** The accelerated expansion of the universe, attributed to dark energy, is studied through supernova surveys and the large-scale structure of the cosmos. Projects like the Dark Energy Survey (DES) seek to understand this mysterious force.

Black Holes and Neutron Stars

Extreme Physics:

- **Event Horizon Telescope (EHT):** The EHT's imaging of black holes tests the limits of general relativity and offers potential insights into quantum gravity.
- **X-ray and Radio Observations:** Telescopes observing X-rays and radio waves from neutron stars and black holes help study their environments and behaviors, testing predictions from string theory and other advanced models.

Future Prospects

Upcoming Missions:

- **James Webb Space Telescope (JWST):** Scheduled for launch, JWST will provide deeper insights into the early universe, star formation, and exoplanets, potentially revealing clues about fundamental physics.
- **Euclid and WFIRST:** These missions will map dark matter and study dark energy with unprecedented precision, contributing to our understanding of the universe's fundamental components.

The Search for Supersymmetric Particles

The search for supersymmetric particles is one of the most significant endeavors in modern particle physics. Supersymmetry extends the Standard Model by positing that every known particle has a superpartner with different spin properties. Discovering these superpartners would provide evidence for SUSY, addressing several unresolved issues in physics. Here's a look at the search for supersymmetric particles.

Theoretical Motivation

Solving the Hierarchy Problem:

- **Stabilizing the Higgs Boson:** Supersymmetry helps to stabilize the mass of the Higgs boson by canceling out large quantum corrections that arise in the Standard Model. Superpartners of the top quark, called stops, are important in this cancellation.

Unification of Forces:

- **Gauge Coupling Unification:** SUSY predicts that the running of the gauge couplings for the electromagnetic, weak, and strong forces converge at a high energy scale, supporting the idea of a grand unified theory (GUT).

Dark Matter Candidate:

- **Lightest Supersymmetric Particle (LSP):** In many SUSY models, the LSP is stable and weakly interacting, making it an excellent candidate for dark matter, which accounts for approximately 27% of the universe's mass-energy content.

Experimental Approaches

Collider Experiments:

- **Large Hadron Collider (LHC):** The LHC at CERN is the world's most powerful particle accelerator, capable of reaching the high energies needed to potentially produce SUSY particles. Experiments like ATLAS and CMS are designed to detect these particles.
- **Collision Data:** Researchers analyze collision data for signs of superpartners, such as missing transverse energy (indicating an undetected LSP) and unique decay patterns distinct from Standard Model processes.

Search Strategies

Missing Energy:

- **LSP Signature:** SUSY particles often decay into the LSP, which is stable and does not interact strongly with ordinary matter, escaping the detector. This results in missing transverse energy, a key indicator in SUSY searches.

Mass Spectrum:

- **Mass Hierarchies:** SUSY models predict a spectrum of masses for superpartners. Experiments search for specific mass ranges based on theoretical predictions, focusing on lighter superpartners that might be within the reach of current collider energies.

Flavor and Charge:

- **Distinct Decay Channels:** Superpartners can decay into unique combinations of Standard Model particles. Identifying these distinct decay channels, especially involving heavy quarks or leptons, can signal SUSY.

Current Status and Challenges

Experimental Results:

- **No Definitive Detection Yet:** Despite extensive searches, no SUSY particles have been conclusively detected at the LHC. This has led to constraints on their masses, pushing the possible mass range for superpartners higher.
- **Refining Models:** The lack of direct evidence has led to refining SUSY models, considering higher mass ranges and alternative decay scenarios.

Technical Challenges:

- **High Energy Requirements:** Producing SUSY particles requires extremely high energies, often beyond current collider capabilities. Future colliders might be necessary to explore these higher energy scales.
- **Background Noise:** Distinguishing potential SUSY signals from the Standard Model background is challenging due to the vast amount of data and similar decay patterns.

Future Prospects

High Luminosity LHC (HL-LHC):

- **Increased Data:** The HL-LHC will significantly increase the collision data, improving the chances of detecting rare SUSY events and allowing more precise measurements of particle properties.

Next-Generation Colliders:

- **Future Circular Collider (FCC):** Proposed future colliders like the FCC aim to achieve even higher energies and luminosities, which could probe deeper into the SUSY parameter space.
- **International Linear Collider (ILC):** The ILC would provide a cleaner experimental environment for precision studies of particles, complementing the discoveries of hadron colliders and enhancing the search for SUSY.

Indirect Searches:

- **Astrophysical Observations:** Experiments designed to detect dark matter particles, such as those looking for weakly interacting massive particles (WIMPs), could provide indirect evidence for SUSY if the LSP is detected.
- **Precision Measurements:** Deviations from Standard Model predictions in precision measurements, such as the magnetic moment of the muon or rare decays, might indicate SUSY effects.

Technological Limitations

Technological limitations are significant hurdles in experimental approaches to testing theories like string theory. Despite advanced equipment and innovative techniques, several factors impede our ability to fully explore and verify these theoretical frameworks.

High Energy Requirements
Particle Accelerators:

- **Current Limits:** The energies required to test many predictions of string theory, including the production of supersymmetric particles or the probing of extra dimensions, are often beyond the reach of current particle accelerators like the Large Hadron Collider (LHC).
- **Future Needs:** Future accelerators, such as the proposed Future Circular Collider (FCC), aim to reach higher energies, but these projects are decades away and require massive financial and technological investments.

Detection Sensitivity
Small Scale Phenomena:

- **Extra Dimensions:** String theory predicts extra dimensions that are compactified to incredibly small scales. Detecting these dimensions directly is challenging because their effects are subtle and require extremely precise measurements.
- **Gravitational Waves:** While gravitational wave detectors like LIGO have made groundbreaking discoveries, detecting waves with the sensitivity needed to observe phenomena related to string theory remains a technological challenge.

Data Processing
Huge Data Volumes:

- **Collider Data:** Experiments at particle colliders produce vast amounts of data. Analyzing this data to find rare events, such as the decay patterns predicted by supersymmetry, requires advanced algorithms and significant computational power.

- **Noise Reduction:** Differentiating between signal and noise in the data is critical. Even with sophisticated data processing techniques, distinguishing potential evidence of new physics from background noise remains difficult.

Precision and Calibration

Instrument Calibration:

- **Detector Precision:** Ensuring that detectors are precisely calibrated to measure particle properties accurately is crucial. Any slight miscalibration can lead to incorrect interpretations of the data.
- **Environmental Factors:** External factors such as temperature fluctuations and electromagnetic interference can affect detector performance, requiring constant monitoring and adjustment.

Financial and Logistical Constraints

Resource Intensive:

- **Cost:** Building and maintaining large-scale experimental facilities, such as particle colliders and gravitational wave detectors, involves enormous financial resources.
- **Collaboration:** These projects often require international collaboration, which can be logistically complex and politically sensitive.

While technological advancements have brought us closer to testing string theory and other advanced models, significant limitations remain. Overcoming these challenges requires further innovations, substantial investment, and international cooperation.

CHAPTER 12: CRITICISMS AND CONTROVERSIES

Theoretical Criticisms

String theory, while a compelling and elegant framework aiming to unify all fundamental forces, faces significant theoretical criticisms. Here are some of the main points of contention.

Lack of Predictive Power
Vast Landscape:

* **String Theory Landscape:** One of the most significant criticisms is the "landscape problem." String theory predicts an enormous number of possible vacuum states—on the order of $10500 10^{500} 10500$. Each vacuum state corresponds to a different possible universe with its own set of physical laws. This vast landscape makes it difficult to extract specific, testable predictions about our universe.

Ambiguity in Predictions:

* **Parameter Tuning:** With so many possible solutions, critics argue that string theory can be adjusted to fit almost any experimental outcome. This flexibility undermines its predictive power and makes it challenging to falsify the theory.

Lack of Experimental Evidence
High Energy Scales:

* **Beyond Current Technology:** Many of string theory's predictions manifest at energy scales far beyond the reach of current particle accelerators. For example, the Planck scale, where quantum gravity effects become significant, is $1019 10^{19} 1019$ times higher than the energies accessible by the Large Hadron Collider (LHC). This makes direct experimental verification nearly impossible with current technology.

Indirect Evidence:

* **Absence of Supersymmetry:** String theory predicts the existence of supersymmetric particles, but despite extensive searches, no such particles have been conclusively detected. This absence of evidence raises doubts about the theory's accuracy.

Mathematical Complexity

Advanced Mathematics:

- **High-Dimensional Spaces:** String theory requires complex mathematical structures, including extra dimensions and advanced concepts from algebraic geometry and topology. Critics argue that this complexity makes the theory less accessible and harder to work with than other models.
- **Intuition and Interpretation:** The mathematics of string theory often lacks intuitive physical interpretations, making it difficult to connect theoretical constructs with observable phenomena.

Philosophical Concerns

Scientific Method:

- **Falsifiability:** Some critics question whether string theory meets the criteria of a scientific theory as defined by Karl Popper, specifically the requirement of falsifiability. Since string theory can accommodate a wide range of outcomes, it's challenging to design experiments that could definitively disprove it.

Anthropic Principle:

- **Anthropic Arguments:** The reliance on the anthropic principle to explain why our universe has the properties it does (among the vast number of possible string vacua) is controversial. Critics argue that this principle shifts the theory from empirical science towards a more philosophical stance, reducing its explanatory power.

Alternative Theories

Competing Models:

- **Loop Quantum Gravity:** Alternative theories, such as loop quantum gravity, offer different approaches to unifying gravity with quantum mechanics. These models often make distinct predictions and are based on different underlying principles, challenging the universality of string theory.
- **Quantum Field Theory Advances:** Developments in quantum field theory and condensed matter physics also provide alternative frameworks for understanding fundamental interactions without requiring the complex constructs of string theory.

Internal Consistency

Mathematical Rigor:

- **Unproven Assumptions:** While string theory is mathematically rich, some of its foundational assumptions remain unproven. The consistency of compactifying extra dimensions, the stability of vacuum states, and the exact nature of brane interactions are areas that require further theoretical work.

Philosophical Debates

String theory, as a leading candidate for a theory of everything, has ignited various philosophical debates. These discussions extend beyond scientific scrutiny, touching upon the nature of scientific theories, the limits of human understanding, and the philosophical implications of a unified description of the universe. Here are some key philosophical debates surrounding string theory.

Falsifiability and Scientific Method
Popper's Criterion:

- **Falsifiability:** Karl Popper, a prominent philosopher of science, argued that a theory must be falsifiable to be considered scientific. Critics of string theory argue that due to its vast landscape of solutions, it is difficult to devise experiments that could definitively falsify it.
- **Predictions and Tests:** While string theory can make specific predictions under certain conditions, the energy scales required to test these predictions are often beyond current experimental capabilities. This raises questions about its status as a scientific theory.

The Nature of Reality
Multiverse Hypothesis:

- **String Landscape:** String theory predicts a multitude of possible vacuum states, each corresponding to a different universe with its own physical laws. This leads to the idea of a multiverse, where our universe is just one of many.
- **Anthropic Principle:** To explain why our universe has the properties it does, some physicists invoke the anthropic principle, suggesting that we observe this particular universe because it allows for the existence of observers like us. Critics argue this principle lacks explanatory power and shifts the theory towards a more philosophical, rather than empirical, realm.

Mathematical Elegance vs. Empirical Evidence
Aesthetic Appeal:

- **Elegance and Beauty:** String theory is often praised for its mathematical elegance and beauty. Proponents argue that its internal consistency and ability to unify all fundamental forces are strong indications of its validity.
- **Empirical Verification:** Philosophers and some physicists caution against equating mathematical elegance with physical truth. They emphasize the importance of empirical evidence and warn that a theory's beauty does not guarantee its correctness.

The Role of Extra Dimensions

Physical Reality:

- **Compactified Dimensions:** String theory requires additional spatial dimensions beyond the familiar three. These extra dimensions are compactified into tiny, complex shapes, making them difficult to observe directly.
- **Metaphysical Implications:** The existence of extra dimensions raises metaphysical questions about the nature of reality. Are these dimensions just mathematical constructs, or do they have a physical existence that we have yet to comprehend?

Epistemological Limits

Human Understanding:

- **Complexity and Comprehensibility:** The mathematical complexity of string theory poses challenges for human understanding. Philosophers debate whether there are fundamental limits to what we can know about the universe.
- **Reductionism vs. Holism:** String theory represents a highly reductionist approach, aiming to explain all physical phenomena through the behavior of strings. Some argue for a more holistic view, suggesting that understanding the universe may require multiple, complementary approaches.

Comparison with Alternative Theories

Loop Quantum Gravity:

- **Competing Frameworks:** Loop quantum gravity (LQG) is a competing theory that also seeks to unify quantum mechanics and general relativity. LQG does not rely on extra dimensions and offers a different perspective on the nature of space and time.
- **Philosophical Implications:** The existence of competing theories like LQG highlights the philosophical debate over theory choice. How should scientists decide between equally compelling but fundamentally different theoretical frameworks?

Theoretical vs. Practical Knowledge

Value of Theoretical Work:

- **Theoretical Contributions:** Despite the lack of direct empirical evidence, string theory has contributed significantly to theoretical physics, leading to advances in mathematics and providing new insights into the nature of black holes and quantum gravity.
- **Practical Application:** Critics argue that the practical impact of string theory on experimental physics has been limited. They question the allocation of resources towards a theory that may not yield testable predictions in the foreseeable future.

String Theory vs. Loop Quantum Gravity

String theory and loop quantum gravity (LQG) are two leading theoretical frameworks aimed at unifying general relativity and quantum mechanics. Each offers a unique approach to understanding the fundamental structure of the universe. Here's a detailed comparison of these two theories.

Basic Principles

String Theory:

- **Fundamental Entities:** String theory posits that the fundamental constituents of the universe are one-dimensional strings rather than point-like particles. These strings can vibrate at different frequencies, with each vibrational mode corresponding to a different particle.
- **Extra Dimensions:** String theory requires additional spatial dimensions beyond the familiar three. In most versions, there are ten dimensions (superstring theory) or eleven dimensions (M-theory).
- **Supersymmetry:** String theory often incorporates supersymmetry, which posits that every particle has a superpartner with different spin properties.

Loop Quantum Gravity:

- **Quantized Space:** LQG proposes that space itself is quantized. Instead of being a continuous entity, space is composed of discrete, finite loops called spin networks.
- **No Extra Dimensions:** LQG does not require extra dimensions; it operates within the familiar four-dimensional spacetime framework.
- **Background Independence:** LQG is background-independent, meaning it does not assume a fixed spacetime background. Instead, the geometry of spacetime is dynamic and emergent.

Mathematical Framework

String Theory:

- **Vibrational Modes:** The mathematics of string theory involves complex equations describing the vibrational modes of strings. These equations are embedded in higher-dimensional spaces, often requiring advanced concepts from algebraic geometry and topology.
- **Gauge Symmetry:** String theory supports various gauge symmetries, which are fundamental to the Standard Model of particle physics.

Loop Quantum Gravity:

- **Spin Networks:** The mathematical framework of LQG is built on spin networks, which represent quantum states of the gravitational field. These networks evolve over time, forming spin foams that describe the quantum geometry of spacetime.
- **Canonical Quantization:** LQG employs canonical quantization techniques to merge quantum mechanics with general relativity, focusing on quantizing the geometry of spacetime itself.

Key Predictions and Implications

String Theory:

- **Unification of Forces:** String theory aims to unify all fundamental forces, including gravity, into a single theoretical framework. It naturally incorporates gravity through the graviton, a vibrational mode of the string.
- **Extra Dimensions and Multiverse:** String theory's requirement for extra dimensions leads to the concept of the multiverse, with a vast number of possible universes corresponding to different solutions in the string landscape.

Loop Quantum Gravity:

- **Discrete Spacetime:** LQG predicts that spacetime is composed of discrete units, leading to potential observable effects at very small scales, such as in black hole interiors or near the Big Bang.
- **Resolution of Singularities:** LQG suggests that singularities, like those in black holes and the Big Bang, are resolved by the quantum nature of spacetime, avoiding the infinities predicted by classical general relativity.

Experimental Prospects

String Theory:

- **High Energy Requirements:** Direct experimental evidence for string theory is challenging due to the high energy scales required to test its predictions, often beyond current or foreseeable collider capabilities.
- **Indirect Evidence:** Researchers look for indirect signs, such as effects of supersymmetric particles or deviations in the behavior of gravity at small scales, but these have not been conclusively observed.

Loop Quantum Gravity:

- **Observable Effects:** LQG's predictions about the quantized nature of spacetime could, in principle, be tested through observations of high-energy astrophysical phenomena or the cosmic microwave background.
- **Quantum Gravity Phenomena:** Potential signatures of LQG might include deviations from classical predictions in extreme conditions, such as near black holes or in the early universe.

Philosophical and Theoretical Challenges
String Theory:

- **Vast Landscape:** The string theory landscape problem, with its multitude of possible solutions, makes it difficult to extract specific, testable predictions about our universe.
- **Mathematical Complexity:** The advanced mathematical structures required for string theory are both a strength and a hindrance, as they can make the theory less accessible and harder to work with.

Loop Quantum Gravity:

- **Completeness:** LQG has been criticized for not yet providing a complete unification of all fundamental forces, focusing primarily on quantizing gravity.
- **Experimental Validation:** While LQG is more background-independent, finding definitive experimental evidence remains a significant challenge.

Complementary Approaches

Some researchers believe that string theory and LQG could be complementary rather than competing. There are efforts to understand how the insights from both frameworks might converge or inform a more comprehensive theory of quantum gravity.

CHAPTER 13: APPLICATIONS OF STRING THEORY

Technology and String Theory

String theory, while primarily a theoretical framework for understanding the fundamental nature of the universe, has intriguing implications for technology. The concepts and mathematical techniques developed in string theory research can influence various technological fields, even though direct applications may still be speculative. Here are some ways string theory impacts technology and inspires innovation.

Advanced Computational Techniques
Simulating Complex Systems:

- **Computational Power:** String theory requires significant computational power to simulate its complex mathematical structures. These demands drive advancements in computational techniques and hardware, which can be applied to other areas such as cryptography, data analysis, and complex simulations in engineering and finance.
- **Algorithm Development:** The algorithms developed to solve string theory equations, such as those involving higher-dimensional geometry and topological field theory, can be adapted to solve problems in computer science and materials science.

Quantum Computing
Inspiration for Quantum Algorithms:

- **Quantum Mechanics Insights:** String theory incorporates deep principles of quantum mechanics. The study of these principles inspires new quantum algorithms and error correction techniques, crucial for the development of practical quantum computers.
- **Entanglement and Information Theory:** Research in string theory, particularly the AdS/CFT correspondence, provides insights into quantum entanglement and information theory, which are foundational for quantum computing.

Materials Science
Topological Materials:

- **Topological Insulators:** Concepts from string theory and related fields, such as topological field theory, contribute to the understanding and development of topological insulators and superconductors. These materials have unique properties that can lead to advances in electronics and quantum devices.

- **Metamaterials:** The mathematical tools used in string theory to describe exotic geometries are applied to design metamaterials with tailored electromagnetic properties, impacting telecommunications, optics, and sensor technology.

Cryptography

Enhanced Security Algorithms:

- **Complex Mathematical Structures:** The advanced mathematics of string theory, including group theory and modular forms, can inspire new cryptographic algorithms that offer enhanced security. These structures provide robust frameworks for developing encryption methods resistant to classical and quantum attacks.

Theoretical Frameworks for New Technologies

Nanotechnology:

- **String-like Models:** String theory's conceptual framework can be adapted to model nanoscale phenomena, providing insights into the behavior of nanomaterials and aiding in the design of nanodevices and nanostructures.
- **Precision Engineering:** Understanding the fundamental interactions at the smallest scales helps develop precision engineering techniques critical for advancing nanotechnology.

Astrophysics and Space Exploration

Advanced Propulsion Concepts:

- **Warp Drives and Wormholes:** While speculative, the theoretical constructs of string theory, such as higher dimensions and the manipulation of spacetime, inspire concepts for advanced propulsion systems like warp drives and wormholes, which could revolutionize space travel.
- **Gravitational Wave Detection:** The understanding of quantum gravity from string theory enhances the design and interpretation of gravitational wave detectors, aiding in the study of cosmic events and the development of space-based observatories.

Medical Imaging and Diagnostics

Imaging Algorithms:

- **Mathematical Techniques:** The sophisticated mathematical techniques developed in string theory can improve imaging algorithms used in medical diagnostics, leading to more accurate and detailed imaging technologies such as MRI and CT scans.

Education and Outreach
STEM Education:

- **Inspiring Students:** String theory, with its profound and captivating concepts, inspires students to pursue careers in STEM fields. The challenging problems and innovative thinking required in string theory research promote critical thinking and problem-solving skills.
- **Interdisciplinary Learning:** The interdisciplinary nature of string theory, combining physics, mathematics, and computer science, encourages a broad educational approach, fostering a new generation of scientists equipped to tackle complex problems.

Cosmological Applications

String theory offers profound insights into cosmology, providing potential explanations for the universe's structure and evolution. Here are some key cosmological applications of string theory.

Early Universe and Inflation
Cosmic Inflation:

- **Brane Inflation:** String theory suggests mechanisms for cosmic inflation, such as brane inflation, where the separation and interaction of higher-dimensional branes drive rapid expansion. This helps explain the uniformity and flatness of the universe.

Structure Formation
Cosmic Strings:

- **Topological Defects:** String theory predicts the existence of cosmic strings, which are one-dimensional topological defects formed during phase transitions in the early universe. These strings could seed the formation of large-scale structures like galaxies.

Dark Matter and Dark Energy
Supersymmetry:

- **Dark Matter Candidates:** String theory often incorporates supersymmetry, predicting stable, weakly interacting particles that could constitute dark matter. The lightest supersymmetric particle (LSP) is a prime candidate for dark matter.
- **Moduli Fields:** The theory also introduces moduli fields, which could contribute to dark energy, explaining the accelerated expansion of the universe.

Black Holes and Singularities

Quantum Gravity:

- **Black Hole Microstates:** String theory provides a statistical description of black hole entropy through the concept of microstates, potentially resolving the information paradox.
- **Resolution of Singularities:** The theory suggests that singularities, such as those at the centers of black holes and the Big Bang, are smoothed out by the quantum nature of strings, avoiding the infinities predicted by classical general relativity.

Multiverse and Extra Dimensions

String Landscape:

- **Multiverse Hypothesis:** String theory's vast landscape of possible vacuum states implies the existence of a multiverse, where different regions of space have different physical laws and constants.
- **Extra Dimensions:** The compactification of extra dimensions influences the properties of our universe, potentially observable through high-energy experiments and cosmological observations.

Future Observations

Gravitational Waves:

- **Cosmic Signatures:** Gravitational wave detectors could potentially observe signatures of cosmic strings or other string-theoretic phenomena, providing indirect evidence for the theory.

Potential Future Applications

String theory, though primarily a theoretical framework, holds potential for a variety of future applications across multiple fields. Here's an exploration of these prospective applications:

Quantum Computing

Quantum Information Theory:

* **Entanglement and Information:** String theory's insights into quantum entanglement and holography can inform the development of quantum computing algorithms and error correction techniques, enhancing the efficiency and reliability of quantum computers.

Quantum Gravity Simulations:

* **Simulating Quantum Systems:** The techniques developed in string theory, particularly those involving the AdS/CFT correspondence, can be used to simulate complex quantum systems, potentially leading to breakthroughs in understanding quantum gravity and developing quantum technologies.

Advanced Materials

Topological Insulators:

* **Exotic Phases of Matter:** Concepts from string theory, such as topological field theories, can aid in the design of new materials with unique electronic properties, like topological insulators and superconductors, revolutionizing electronics and quantum devices.

Metamaterials:

* **Customized Properties:** The mathematical tools used in string theory to describe complex geometries can be applied to design metamaterials with tailored electromagnetic properties, impacting telecommunications, optics, and sensor technology.

High-Energy Physics

Collider Experiments:

* **Discovering New Particles:** Future high-energy colliders, inspired by string theory predictions, could uncover supersymmetric particles, providing direct evidence for the theory and leading to a deeper understanding of fundamental physics.

Probing Extra Dimensions:

- **Experimental Verification:** Advanced collider experiments could test the existence of extra dimensions predicted by string theory, providing crucial insights into the nature of our universe.

Astrophysics and Cosmology

Gravitational Wave Astronomy:

- **Detecting Cosmic Strings:** Gravitational wave detectors might observe signatures of cosmic strings, a phenomenon predicted by string theory, offering evidence for the theory and enriching our understanding of the early universe.

Black Hole Physics:

- **Quantum Aspects of Black Holes:** String theory's approach to black hole microstates and entropy can lead to new methods for studying black holes, potentially resolving paradoxes like the information loss problem.

Medical and Biological Applications

Medical Imaging:

- **Improved Algorithms:** The sophisticated mathematical techniques developed in string theory can enhance imaging algorithms used in medical diagnostics, leading to more accurate and detailed imaging technologies such as MRI and CT scans.

Nanotechnology:

- **Nanoscale Modeling:** String theory's framework can be adapted to model nanoscale phenomena, providing insights into the behavior of nanomaterials and aiding in the design of nanodevices and nanostructures.

Computational Advances

Algorithm Development:

- **Enhanced Data Analysis:** The complex algorithms developed for string theory calculations can be adapted to improve data analysis techniques in various fields, from finance to climate science, making sense of large and complex datasets.

Artificial Intelligence:

- **Machine Learning:** Insights from string theory might inform the development of new machine learning models and artificial intelligence systems, particularly in areas requiring the modeling of high-dimensional data.

Fundamental Physics and Beyond

Unifying Theories:

- **Theory of Everything:** String theory continues to aim for a unified theory of all fundamental forces, potentially leading to a comprehensive understanding of the universe's underlying principles and inspiring new technological advancements based on this profound knowledge.

Philosophical Implications:

- **Nature of Reality:** Beyond practical applications, string theory's exploration of higher dimensions and the fabric of spacetime could revolutionize our philosophical understanding of reality, influencing fields such as metaphysics and epistemology.

While many applications of string theory remain speculative, its profound insights into quantum mechanics, advanced materials, high-energy physics, and other areas hold significant potential for future technological advancements and a deeper understanding of the universe. The ongoing development and exploration of string theory continue to inspire and drive innovation across multiple domains.

String Theory in Other Scientific Fields

String theory has influenced and contributed to various other scientific fields. Its mathematical structures, concepts, and techniques have found applications beyond theoretical physics, impacting diverse areas of science. Here are some examples of string theory's influence in other scientific fields:

Mathematics

Advanced Mathematical Tools:

- **Algebraic Geometry and Topology:** String theory has driven advances in algebraic geometry and topology. Techniques developed in these areas to handle complex manifolds and higher-dimensional spaces have led to new insights and theorems, enhancing our understanding of mathematical structures.

- **Mirror Symmetry:** Mirror symmetry, a concept from string theory, has provided profound insights into the relationship between different Calabi-Yau manifolds. This has led to significant progress in enumerative geometry and the understanding of complex structures.

Quantum Information Theory
Entanglement and Holography:

- **AdS/CFT Correspondence:** The AdS/CFT correspondence, a central result in string theory, has been instrumental in understanding quantum entanglement and information theory. This duality provides a framework for studying the entanglement structure of quantum states and has applications in quantum computing and quantum error correction.
- **Quantum Entropy:** Insights from string theory have contributed to the understanding of entropy and information flow in quantum systems, influencing research in quantum information theory.

Condensed Matter Physics
Topological Phases of Matter:

- **Topological Insulators:** Concepts from string theory, such as topological field theories, have been applied to study and understand topological insulators and superconductors. These materials exhibit unique electronic properties with potential applications in quantum computing and advanced electronics.
- **Quantum Criticality:** String theory methods have been used to study phase transitions and critical phenomena in condensed matter systems, providing new ways to model and understand these complex behaviors.

Astrophysics and Cosmology
Early Universe and Inflation:

- **Cosmic Inflation:** String theory offers potential mechanisms for cosmic inflation, such as brane inflation. These models help explain the rapid expansion of the early universe and its uniformity.
- **Structure Formation:** The theory predicts the existence of cosmic strings, which could have played a role in the formation of large-scale structures in the universe, such as galaxies and clusters.

Black Hole Physics:

- **Information Paradox:** String theory's approach to black hole microstates and entropy has provided new insights into the black hole information paradox, suggesting that information is preserved in black hole evaporation.

- **Hawking Radiation:** Understanding quantum aspects of black holes through string theory has implications for studying Hawking radiation and the thermodynamics of black holes.

Biological Systems and Medical Applications

Modeling Complex Systems:

- **Protein Folding:** Techniques from string theory have been applied to model the folding of proteins and other biomolecules, providing insights into their behavior and interactions at a molecular level.
- **Medical Imaging:** The sophisticated mathematical tools developed in string theory have potential applications in improving medical imaging techniques, such as MRI and CT scans, leading to more accurate and detailed diagnostics.

Climate Science and Environmental Modeling

Data Analysis Techniques:

- **Complex Systems:** The advanced data analysis and computational techniques developed in string theory research can be applied to model and understand complex environmental systems, such as climate dynamics and ecological interactions.
- **Predictive Models:** These techniques help create more accurate predictive models for studying climate change and its impacts, aiding in the development of mitigation strategies.

Neuroscience

Neural Networks:

- **Modeling Brain Activity:** The mathematical frameworks and computational techniques from string theory can be used to model neural networks and brain activity, contributing to a better understanding of cognitive processes and neural dynamics.
- **Complex Systems Analysis:** String theory's approach to handling complex, high-dimensional systems provides tools for analyzing the intricate connections and interactions within the brain.

Materials Science

Designing New Materials:

- **Metamaterials:** String theory's mathematical structures can be applied to design metamaterials with customized properties, impacting fields such as telecommunications, optics, and sensor technology.
- **Nanotechnology:** The principles of string theory can aid in modeling nanoscale phenomena, providing insights into the behavior of nanomaterials and aiding in the design of nanodevices and nanostructures.

String theory's influence extends beyond theoretical physics into various scientific fields. Its advanced mathematical techniques, conceptual frameworks, and computational methods have found applications in mathematics, quantum information theory, condensed matter physics, astrophysics, biology, climate science, neuroscience, and materials science. These interdisciplinary applications demonstrate the far-reaching impact of string theory on our understanding of the natural world and the development of new technologies.

CHAPTER 14: THE FUTURE OF STRING THEORY

Next Steps in Research

The future of string theory lies in both theoretical advancements and experimental validation. Researchers are actively exploring new ideas and refining existing frameworks to bridge the gap between string theory and observable phenomena. Here are the next steps in string theory research:

Refining Mathematical Frameworks

Dualities and Correspondences:

- **AdS/CFT Correspondence:** Continuing to explore the AdS/CFT correspondence will deepen our understanding of the relationship between gravity and quantum field theories. This duality provides a powerful tool for studying black holes and strongly coupled systems.
- **Other Dualities:** Researchers will further investigate other dualities, such as T-duality and S-duality, to uncover deeper connections within string theory and between different physical theories.

Higher-Dimensional Models:

- **Compactification:** Refining models of compactification, which describe how extra dimensions are curled up, remains a key area of research. Understanding the shapes and sizes of these dimensions is crucial for making string theory predictions that match our four-dimensional universe.
- **Calabi-Yau Manifolds:** Studying the properties of Calabi-Yau manifolds, the mathematical spaces where extra dimensions may reside, will help in developing more precise models.

Bridging Theory and Experiment

Phenomenology:

- **Connecting to Observables:** String phenomenology focuses on connecting string theory to observable phenomena, such as particle masses and coupling constants. This involves translating the abstract mathematics of string theory into predictions that can be tested in experiments.
- **Supersymmetry Searches:** Researchers will continue to search for evidence of supersymmetry in high-energy particle collisions, such as those conducted at the Large Hadron Collider (LHC). Detecting superpartners would provide strong support for string theory.

Gravitational Wave Observations:

- **Cosmic Strings:** Gravitational wave detectors like LIGO and Virgo may observe signals from cosmic strings, hypothetical one-dimensional defects predicted by string theory. Detecting these would offer indirect evidence for the theory.
- **Black Hole Physics:** Observing the detailed properties of black holes and their mergers can provide insights into quantum gravity and potentially confirm string theory predictions about black hole entropy and information retention.

Exploring Cosmological Implications

Early Universe:

- **Inflation Models:** Developing and refining string-theoretic models of cosmic inflation will help explain the rapid expansion of the early universe and its large-scale structure. Brane inflation, in particular, is a promising area of research.
- **Dark Matter and Dark Energy:** Investigating the role of string-theoretic particles, such as axions or the lightest supersymmetric particles, in explaining dark matter and dark energy will be crucial. These components dominate the universe's mass-energy content and understanding them is essential.

Advancing Computational Techniques

Simulation and Data Analysis:

- **Advanced Algorithms:** Developing sophisticated algorithms to simulate string-theoretic models and analyze vast datasets from particle colliders and astrophysical observations is essential. These tools will help bridge the gap between theoretical predictions and experimental data.
- **Machine Learning:** Incorporating machine learning techniques to identify patterns and make predictions in complex data sets will enhance the ability to test string theory against empirical evidence.

Collaborative and Interdisciplinary Efforts

International Collaboration:

- **Global Projects:** Large-scale projects like the proposed Future Circular Collider (FCC) and space-based gravitational wave detectors will require international collaboration. These efforts will push the boundaries of our experimental capabilities and provide opportunities to test string theory in new regimes.

- **Interdisciplinary Research:** Collaborating with mathematicians, computer scientists, and other physicists will bring fresh perspectives and tools to tackle the complex problems in string theory.

Competing Theories and Approaches to String Theory

While string theory remains a leading candidate for a unified theory of all fundamental forces, several competing theories and approaches also seek to reconcile general relativity and quantum mechanics. Here's an exploration of these alternatives and their unique perspectives.

Loop Quantum Gravity (LQG)

Fundamental Concepts:

- **Quantized Space:** LQG proposes that space itself is quantized, composed of discrete loops called spin networks.
- **Background Independence:** Unlike string theory, LQG does not assume a fixed spacetime background. The geometry of spacetime is dynamic and emergent from the quantum properties of these loops.

Key Features:

- **Resolution of Singularities:** LQG aims to resolve singularities, such as those in black holes and the Big Bang, by using the quantum nature of spacetime.
- **No Extra Dimensions:** LQG operates within the familiar four-dimensional spacetime framework, avoiding the need for extra dimensions.

Challenges:

- **Unification of Forces:** LQG primarily focuses on quantizing gravity and does not yet provide a comprehensive unification of all fundamental forces.

Causal Dynamical Triangulations (CDT)

Fundamental Concepts:

- **Discrete Spacetime:** CDT models spacetime as a collection of simplexes (triangular building blocks) that evolve dynamically.
- **Causality:** CDT respects causality, ensuring that the sequence of events follows a cause-and-effect relationship.

Key Features:

- **Emergent Continuum:** CDT demonstrates how a continuous spacetime can emerge from discrete building blocks.
- **Simplicity:** The model's simplicity allows for straightforward numerical simulations.

Challenges:

- **Complexity of Interactions:** CDT's discrete nature can make modeling complex interactions challenging, and it does not yet fully integrate all forces.

Asymptotic Safety

Fundamental Concepts:

- **Renormalization Group:** This approach suggests that gravity becomes "safe" at high energies due to a fixed point in the renormalization group flow, where physical quantities remain finite and well-behaved.
- **Predictive Power:** If gravity is asymptotically safe, it could be described by a finite number of parameters, making the theory predictive.

Key Features:

- **Non-Perturbative:** Asymptotic safety relies on non-perturbative methods, allowing for a well-defined quantum theory of gravity without divergences.

Challenges:

- **Empirical Testing:** The theory's reliance on high-energy behavior makes experimental verification challenging.

Emergent Gravity

Fundamental Concepts:

- **Emergence from Information:** This approach posits that gravity is not a fundamental force but emerges from the statistical behavior of microscopic degrees of freedom, similar to how thermodynamics emerges from molecular interactions.

Key Features:

- **Entropic Gravity:** Proposed by Erik Verlinde, entropic gravity suggests that gravity arises from changes in entropy associated with the positions of masses.
- **Holographic Principle:** This approach often employs the holographic principle, suggesting that all information within a volume can be described by a theory on the boundary of that volume.

Challenges:

- **Experimental Verification:** The emergent nature of gravity makes direct testing difficult, and the theory's predictions need further development.

Non-Commutative Geometry

Fundamental Concepts:

- **Space and Time as Operators:** In non-commutative geometry, the coordinates of space and time do not commute, meaning their multiplication depends on the order in which they are applied.
- **Quantum Groups:** This approach uses quantum groups to describe the symmetry of spacetime, leading to modified equations of motion.

Key Features:

- **Modifications to Standard Physics:** Non-commutative geometry can lead to small corrections to the Standard Model and general relativity, potentially observable in high-precision experiments.

Challenges:

- **Mathematical Complexity:** The non-commutative nature of spacetime introduces significant mathematical complexity, making the theory challenging to work with and interpret.

Twistor Theory

Fundamental Concepts:

- **Twistors as Fundamental Objects:** Twistor theory, developed by Roger Penrose, proposes that the fundamental objects are not points in spacetime but twistors, which encode information about light rays.
- **Complex Geometry:** This approach uses complex geometry to describe the properties of spacetime.

Key Features:

- **Simplification of Equations:** Twistor theory can simplify certain equations in general relativity and quantum field theory, providing new insights and solutions.
- **Connections to String Theory:** There are connections between twistor theory and string theory, particularly in the context of scattering amplitudes.

Challenges:

- **Integration with Quantum Mechanics:** Fully integrating twistor theory with quantum mechanics and developing a comprehensive theory remains an ongoing challenge.

Quantum Graphity
Fundamental Concepts:

- **Graph-Based Spacetime:** Quantum graphity models spacetime as a dynamic graph, with nodes representing quantum states and edges representing interactions.
- **Emergent Geometry:** The geometry of spacetime emerges from the changing connections between nodes.

Key Features:

- **Discrete Model:** This approach provides a discrete model of spacetime, which can be useful for numerical simulations.
- **Dynamic Evolution:** The graph evolves over time, potentially offering insights into the early universe and black hole dynamics.

Challenges:

- **Physical Interpretation:** Interpreting the physical meaning of the graph's evolution and connecting it to observable phenomena is complex.

So while string theory remains a leading framework for unifying fundamental forces, several competing theories and approaches offer unique perspectives and solutions to the challenges of quantum gravity. Each theory has its strengths and challenges, contributing to a diverse and dynamic field of research.

Potential Breakthroughs

The field of string theory is ripe with potential breakthroughs that could revolutionize our understanding of the universe. Here are some of the most promising areas where significant advancements might occur:

Discovery of Supersymmetric Particles

High-Energy Colliders:

- **Detection of Superpartners:** One of the most anticipated breakthroughs would be the discovery of supersymmetric particles, such as the neutralino or the gluino, at high-energy colliders like the Large Hadron Collider (LHC). Finding these particles would provide strong evidence for supersymmetry, a key component of string theory.
- **Impact on Dark Matter:** The detection of the lightest supersymmetric particle (LSP) could solve the mystery of dark matter, as these particles are prime candidates for dark matter constituents.

Evidence of Extra Dimensions

Collider Experiments:

- **Signatures of Extra Dimensions:** Observing phenomena that indicate the existence of extra spatial dimensions, such as deviations from the inverse-square law of gravity at small scales or the production of Kaluza-Klein particles, would support string theory's prediction of additional dimensions.
- **Resonances in Particle Collisions:** High-energy collisions might produce resonances or missing energy signatures that point to the existence of extra dimensions.

Gravitational Waves and Cosmic Strings

Astrophysical Observations:

- **Detection of Cosmic Strings:** Gravitational wave detectors like LIGO and Virgo could potentially detect the unique signals of cosmic strings, one-dimensional defects predicted by string theory. Observing these signals would provide compelling evidence for the theory.
- **Gravitational Wave Signatures:** Detailed analysis of gravitational wave data could reveal signatures of extra dimensions or other string-theoretic phenomena.

Black Hole Physics

Information Paradox and Microstates:

- **Hawking Radiation Studies:** Advances in understanding Hawking radiation and the black hole information paradox could provide crucial insights. If string theory correctly describes black hole microstates, it might resolve these longstanding issues in theoretical physics.
- **Black Hole Entropy:** Verifying the string-theoretic calculation of black hole entropy through observational or experimental means would be a significant breakthrough.

Quantum Gravity and Holography

AdS/CFT Correspondence:

- **Quantum Gravity Insights:** Further development of the AdS/CFT correspondence could provide a complete quantum description of gravity, bridging the gap between general relativity and quantum mechanics.
- **Applications in Condensed Matter:** Applying the principles of AdS/CFT to solve problems in condensed matter physics, such as high-temperature superconductivity, could demonstrate the practical utility of string theory beyond fundamental physics.

Advancements in Mathematical Frameworks

Refining Mathematical Tools:

- **Calabi-Yau Manifolds:** Better understanding and classification of Calabi-Yau manifolds, the shapes in which extra dimensions might be compactified, could lead to more accurate physical predictions.
- **Non-Perturbative Methods:** Developing non-perturbative methods to solve string theory equations could overcome current limitations and provide exact solutions to longstanding problems.

Unifying Theories and Experimental Techniques

Interdisciplinary Research:

- **Integration with Quantum Information:** Integrating string theory with quantum information theory could lead to new insights into quantum gravity, entanglement, and the nature of spacetime.
- **Cross-Disciplinary Innovations:** Collaborations between string theorists, mathematicians, and experimental physicists could result in innovative experimental techniques to test string-theoretic predictions.

New Particle Physics Discoveries

Beyond the Standard Model:

- **Unexpected Particles:** Discovering new particles or forces that do not fit within the Standard Model but are predicted by string theory could open new avenues for research and confirm aspects of the theory.
- **String-Theoretic Corrections:** Detecting subtle corrections to the Standard Model predictions, such as slight deviations in particle interactions, would provide indirect evidence for string theory.

Cosmological Implications

Early Universe and Inflation:

- **String-Theoretic Models of Inflation:** Developing and validating models of cosmic inflation based on string theory, such as brane inflation, could provide a consistent explanation for the early universe's rapid expansion and large-scale structure.
- **Observational Data Correlation:** Correlating string-theoretic predictions with precise cosmological data from missions like the James Webb Space Telescope (JWST) and the Planck satellite could validate these models.

Potential breakthroughs in string theory span a wide range of areas, from high-energy particle physics and astrophysical observations to advancements in mathematical frameworks and interdisciplinary research.

Unification with Other Theories

String theory's ultimate goal is to provide a unified framework for all fundamental forces and particles in the universe. Unification with other theories is critical to achieving this goal. Let's look at how string theory aims to integrate with various fundamental theories and the potential pathways for unification.

Integration with Quantum Field Theory

Gauge Symmetry:

- **Incorporation of the Standard Model:** String theory naturally incorporates the principles of quantum field theory and gauge symmetries. The vibrational modes of strings correspond to particles in the Standard Model, such as quarks, leptons, and gauge bosons.
- **Extra Gauge Groups:** Beyond the Standard Model, string theory predicts additional gauge groups that could explain phenomena not covered by current physics, potentially leading to the discovery of new particles.

Supersymmetry:

- **Extension of Symmetries:** String theory includes supersymmetry (SUSY), which posits a symmetry between bosons and fermions. Integrating SUSY with the Standard Model helps address theoretical issues such as the hierarchy problem and provides candidates for dark matter.

Unification with General Relativity

Quantum Gravity:

- **Graviton as a Vibrational Mode:** String theory uniquely incorporates gravity by predicting the existence of the graviton, a quantum particle that mediates the gravitational force, as a vibrational mode of a closed string.
- **Smooth Singularities:** String theory provides mechanisms to smooth out the singularities present in general relativity, such as those found in black holes and the Big Bang, by describing them with string dynamics and higher-dimensional objects.

Relationship with Loop Quantum Gravity (LQG)

Complementary Approaches:

- **Background Independence:** While string theory typically assumes a fixed spacetime background, LQG does not. Finding a way to reconcile these approaches could provide a more comprehensive theory of quantum gravity.
- **Spin Networks and Strings:** Researchers explore the potential for integrating spin networks from LQG with the string framework, potentially revealing deeper insights into the fabric of spacetime.

Holography and AdS/CFT Correspondence

Holographic Principle:

- **AdS/CFT Duality:** The AdS/CFT correspondence is a major breakthrough, demonstrating a duality between a gravity theory in AdS space (anti-de Sitter) and a conformal field theory on its boundary. This duality has profound implications for understanding quantum gravity and gauge theories.
- **Emergent Spacetime:** The holographic principle suggests that spacetime and gravity can emerge from lower-dimensional quantum field theories, offering a pathway to unify these concepts within string theory.

Unification with Cosmological Theories

Inflation and String Cosmology:

- **Brane Inflation Models:** String theory proposes models of cosmic inflation involving the dynamics of branes. These models can potentially explain the rapid expansion of the early universe and align with cosmological observations.
- **String Landscape:** The concept of the string landscape, with its multitude of possible vacuum states, offers explanations for the observed values of physical constants and the structure of our universe through the anthropic principle.

Interdisciplinary Connections

Quantum Information Theory:

- **Entanglement and Geometry:** String theory's insights into quantum entanglement and the geometry of spacetime can inform and benefit from developments in quantum information theory. Concepts like holography are particularly relevant for understanding the quantum nature of information.
- **AdS/CFT in Quantum Computing:** Techniques from AdS/CFT correspondence are being applied to solve complex problems in quantum computing, demonstrating interdisciplinary potential.

Condensed Matter Physics:

- **Emergent Phenomena:** Methods from string theory, particularly those involving topological field theories, are used to understand emergent phenomena in condensed matter systems, such as high-temperature superconductivity and quantum phase transitions.
- **Holographic Dualities:** Applying holographic dualities to condensed matter physics offers new ways to study strongly correlated systems and could lead to new materials with exotic properties.

Experimental Prospects

Future Colliders and Detectors:

- **Testing Predictions:** High-energy experiments at future colliders, such as the proposed Future Circular Collider (FCC), could test predictions of string theory, such as supersymmetric particles or extra dimensions.
- **Gravitational Wave Observations:** Detectors like LIGO and Virgo may observe signals of cosmic strings or other phenomena predicted by string theory, providing indirect evidence for the theory.

Implications for Our Understanding of the Universe

The future of string theory holds profound implications for our understanding of the universe. As research progresses, string theory could revolutionize our concepts of space, time, and fundamental interactions.

Unification of Forces

A Single Framework:

- **Fundamental Forces:** String theory aims to unify all fundamental forces —gravity, electromagnetism, and the strong and weak nuclear forces—into a single theoretical framework. This unification would resolve the longstanding issue of integrating gravity with quantum mechanics.
- **Graviton:** The prediction of the graviton, a particle that mediates gravitational interactions, naturally emerges from string theory, potentially providing a quantum description of gravity.

Nature of Particles

Beyond Point Particles:

- **Strings as Fundamental Entities:** Unlike traditional particle physics, which models particles as point-like, string theory describes them as one-dimensional strings. These strings can vibrate at different frequencies, with each mode corresponding to a different particle.
- **Diversity of Particles:** This framework can explain the diversity of particles in the Standard Model and potentially predict new particles, including those that may account for dark matter.

Extra Dimensions

Beyond the Visible Universe:

- **Additional Dimensions:** String theory posits the existence of extra spatial dimensions beyond the familiar three. These dimensions are compactified, meaning they are curled up at incredibly small scales, making them difficult to detect directly.
- **Impact on Physical Laws:** The shape and size of these extra dimensions influence the properties of particles and forces, offering explanations for various physical constants and phenomena.

Cosmology and the Early Universe

Inflation and Structure Formation:

- **Cosmic Inflation:** String theory provides mechanisms for cosmic inflation, a rapid expansion of the early universe. This helps explain the uniformity and large-scale structure of the cosmos.

- **Cosmic Strings:** Predicted by string theory, cosmic strings are one-dimensional defects that could have played a role in structure formation in the universe. Detecting these strings would provide evidence for the theory.

Black Holes and Singularities

Quantum Gravity Effects:

- **Black Hole Information Paradox:** String theory offers potential resolutions to the black hole information paradox, suggesting that information is not lost in black holes but is encoded in subtle ways.
- **Microstates:** The theory provides a framework for understanding black hole entropy through the concept of microstates, potentially resolving singularities predicted by general relativity.

The Holographic Principle

Spacetime and Information:

- **AdS/CFT Correspondence:** The AdS/CFT correspondence, a cornerstone of string theory, implies that a higher-dimensional theory of gravity can be equivalent to a lower-dimensional quantum field theory. This suggests that our universe might be describable as a hologram, with all information encoded on a lower-dimensional boundary.
- **Quantum Entanglement:** This principle has profound implications for understanding quantum entanglement and the nature of spacetime, suggesting that spacetime itself could emerge from more fundamental quantum informational principles.

Dark Matter and Dark Energy

New Candidates:

- **Supersymmetric Particles:** String theory often incorporates supersymmetry, predicting new particles that could constitute dark matter. Detecting these particles would provide significant support for the theory.
- **Moduli Fields:** The theory also introduces moduli fields, which could contribute to dark energy and explain the accelerated expansion of the universe.

Fundamental Constants and Anthropic Principle

Explaining the Universe's Properties:

- **String Landscape:** The multitude of possible vacuum states in string theory, known as the string landscape, could explain why physical constants have the values they do. This supports the anthropic principle, which posits that we observe certain values because they allow for the existence of observers like us.

Future Experimental Validation

Testing Predictions:

- **High-Energy Physics:** Advances in particle accelerators, such as the proposed Future Circular Collider (FCC), could test predictions of string theory, such as the existence of supersymmetric particles or evidence of extra dimensions.
- **Gravitational Waves:** Observations from gravitational wave detectors like LIGO and Virgo may reveal signatures of cosmic strings or other phenomena predicted by string theory.

The future of string theory promises to deepen our understanding of the universe, potentially unifying all fundamental forces, revealing the nature of dark matter and dark energy, and providing a comprehensive framework for quantum gravity and cosmology. As research advances, the implications for both theoretical and experimental physics are vast and profound.

APPENDIX

Terms and Definitions

These terms cover a broad range of concepts and techniques central to the understanding and study of string theory.

1. **String Theory**: A theoretical framework in which point-like particles are replaced by one-dimensional objects called strings.
2. **Superstring Theory**: A version of string theory that includes supersymmetry, positing that each particle has a superpartner.
3. **M-Theory**: An extension of string theory in 11 dimensions, unifying the five different superstring theories.
4. **Brane**: Multidimensional objects in string theory on which open strings can end.
5. **D-Brane**: A type of brane where open strings can attach, with different dimensions (D0, D1, etc.).
6. **Supersymmetry (SUSY)**: A theoretical symmetry between bosons and fermions.
7. **Graviton**: The hypothetical quantum particle that mediates the force of gravity, represented as a closed string in string theory.
8. **Calabi-Yau Manifold**: A complex, compact, six-dimensional shape used to describe the extra dimensions in string theory.
9. **Compactification**: The process of curling up extra dimensions so they are small and unobservable at low energies.
10. **AdS/CFT Correspondence**: A duality between a type of string theory in anti-de Sitter space and a conformal field theory on its boundary.
11. **Bosonic String Theory**: The original version of string theory that only includes bosons and is defined in 26 dimensions.
12. **Type I String Theory**: A superstring theory that includes both open and closed strings and has one type of supersymmetry.
13. **Type IIA String Theory**: A superstring theory with non-chiral (non-handled) fermions in 10 dimensions.
14. **Type IIB String Theory**: A superstring theory with chiral (handled) fermions in 10 dimensions.
15. **Heterotic String Theory**: A combination of two different string theories, producing a consistent theory in 10 dimensions.
16. **String Landscape**: The vast number of possible vacuum states in string theory.
17. **Quantum Gravity**: The field of theoretical physics that seeks to describe gravity according to the principles of quantum mechanics.
18. **Planck Scale**: The energy scale at which quantum effects of gravity become significant, around 10^{19} GeV.
19. **String Coupling Constant**: A parameter that determines the strength of the interaction between strings.

20. **T-Duality**: A symmetry in string theory that relates two theories compactified on circles of radius R and $1/R$.

21. **S-Duality**: A symmetry that relates strong coupling to weak coupling in string theory.

22. **U-Duality**: A comprehensive duality combining both T-duality and S-duality in string theory.

23. **Conformal Field Theory (CFT)**: A quantum field theory that is invariant under conformal transformations.

24. **Worldsheet**: The two-dimensional surface swept out by a string moving through spacetime.

25. **Polyakov Action**: The action describing the dynamics of a string, formulated in terms of the worldsheet.

26. **Central Charge**: A parameter in conformal field theory that measures the number of degrees of freedom.

27. **Moduli Space**: The parameter space of all possible shapes of the extra dimensions in string theory.

28. **Gauge Symmetry**: A type of symmetry that involves transformations of the fields in a quantum field theory.

29. **Holographic Principle**: The idea that all of the information contained within a volume of space can be described by a theory on its boundary.

30. **Cosmic String**: A hypothetical one-dimensional topological defect predicted by string theory.

31. **Kaluza-Klein Theory**: A framework that extends general relativity to higher dimensions.

32. **Anti-de Sitter (AdS) Space**: A spacetime with constant negative curvature, used in the AdS/CFT correspondence.

33. **BPS State**: A stable state in string theory that preserves some supersymmetry.

34. **NS5-Brane**: A type of brane with five spatial dimensions, playing a role in non-perturbative string theory effects.

35. **String Tension**: The energy per unit length of a string, determining its dynamics.

36. **F-Theory**: An extension of string theory that includes varying coupling constants, formulated in 12 dimensions.

37. **Duality**: A relationship between two theories that describe the same physical phenomena.

38. **Winding Mode**: A state in string theory where the string wraps around a compact dimension.

39. **Dilaton**: A hypothetical scalar field in string theory associated with the string coupling constant.

40. **Axion**: A hypothetical particle arising in string theory, potentially a component of dark matter.

41. **Moduli Stabilization**: The process of fixing the values of moduli fields to obtain a stable vacuum in string theory.

42. **Green-Schwarz Mechanism**: A method to cancel anomalies in certain string theories.

43. **Superstring Action**: The action that describes the dynamics of superstrings, including fermionic and bosonic components.

44. **Dirac-Born-Infeld (DBI) Action**: An action describing the dynamics of D-branes in string theory.

45. **Chern-Simons Theory**: A topological field theory used in string theory and M-theory.

46. **Matrix Theory**: A non-perturbative formulation of M-theory using matrices.

47. **Little String Theory**: A non-gravitational theory that describes certain limits of string theory.

48. **String Field Theory**: A formalism that describes string theory in terms of field theory, with strings as the fundamental fields.

49. **NS-NS Sector**: The sector of string theory containing states with Neveu-Schwarz boundary conditions.

50. **R-R Sector**: The sector of string theory containing states with Ramond-Ramond boundary conditions.

51. **Worldsheet Supersymmetry**: Supersymmetry defined on the worldsheet of a string.

52. **Scherk-Schwarz Mechanism**: A method to break supersymmetry by compactifying on a circle with a twist.

53. **Conifold Transition**: A process in string theory where the topology of extra dimensions changes through a singularity.

54. **Gromov-Witten Invariants**: Mathematical tools used to count curves on Calabi-Yau manifolds.

55. **Mirror Symmetry**: A duality between different Calabi-Yau manifolds in string theory.

56. **M-Theory Membrane**: A two-dimensional surface in M-theory, a higher-dimensional analogue of strings.

57. **Fuzzball Proposal**: A model suggesting that black hole microstates can be described as "fuzzballs" in string theory.

58. **Type IIA/B Orientifold**: A construction in string theory involving orientifold planes, leading to specific types of string models.

59. **Strominger-Vafa Calculation**: A method to calculate the entropy of certain black holes using string theory.

60. **String Gas Cosmology**: A model of the early universe based on string theory, with strings and branes playing a key role.

61. **Black Hole Microstates**: Different quantum states in string theory that correspond to a single macroscopic black hole.

62. **Heterotic String**: A type of string theory that combines features of both bosonic strings and superstrings.

63. **Twistor String Theory**: A formulation of string theory using twistor space, providing a different perspective on scattering amplitudes.

64. **Kähler Moduli**: Parameters that describe the shape and size of the extra dimensions in a Calabi-Yau manifold.

65. **Flux Compactification**: A method to stabilize the shape of extra dimensions using background fluxes in string theory.

String Theory Timeline

Here is a timeline outlining the significant milestones and developments in the history of string theory:

1960s
1968:

- **Veneziano Amplitude:** Gabriele Veneziano discovers a mathematical formula, the Veneziano amplitude, that describes the scattering of particles in a way that suggests a connection to strings.

1970s
1970:

- **Dual Resonance Model:** Leonard Susskind, Holger Bech Nielsen, and Yoichiro Nambu independently realize that the Veneziano amplitude can be derived from a theory of one-dimensional strings, marking the birth of string theory.

1971:

- **Fermions and Supersymmetry:** Pierre Ramond, John Schwarz, and André Neveu extend string theory to include fermions, leading to the development of supersymmetry in string theory (superstrings).

1974:

- **Graviton Discovery:** John Schwarz and Joel Scherk demonstrate that string theory naturally includes a particle with the properties of the graviton, suggesting string theory could be a theory of quantum gravity.

1976:

- **Bosonic String Theory:** Michael Green and John Schwarz show that bosonic string theory (without supersymmetry) is inconsistent due to anomalies.

1980s
1984:

- **Anomaly Cancellation:** Michael Green and John Schwarz discover the Green-Schwarz mechanism, showing that anomalies in superstring theory can be canceled, making the theory consistent. This discovery revitalizes interest in string theory.

1985:

- **Heterotic String Theory:** David Gross, Jeffrey Harvey, Emil Martinec, and Ryan Rohm develop heterotic string theory, which combines elements of bosonic and superstring theories.

1986:

- **Calabi-Yau Manifolds:** Philip Candelas, Gary Horowitz, Andrew Strominger, and Edward Witten show how extra dimensions in string theory can be compactified using Calabi-Yau manifolds.

1987:

- **First Superstring Revolution:** The discoveries of anomaly cancellation, heterotic string theory, and compactification lead to the first superstring revolution, establishing string theory as a major research area in theoretical physics.

1990s
1995:

- **M-Theory:** Edward Witten proposes M-theory, which unifies the five different superstring theories in 11 dimensions. This leads to the second superstring revolution and the understanding that these theories are different limits of a single underlying theory.
- **D-Branes:** Joseph Polchinski discovers D-branes, objects on which strings can end, playing a big part in non-perturbative string theory.

1997:

- **AdS/CFT Correspondence:** Juan Maldacena proposes the AdS/CFT correspondence, suggesting a duality between string theory in anti-de Sitter space and a conformal field theory on its boundary. This provides a framework for studying quantum gravity and strongly coupled systems.

2000s
2000:

- **Large Extra Dimensions:** Nima Arkani-Hamed, Savas Dimopoulos, and Gia Dvali propose models with large extra dimensions, suggesting that extra dimensions could be as large as a millimeter, which might be testable in experiments.

2003:

- **Landscape Problem:** The realization of the string theory landscape, with a vast number of possible vacuum states, leads to discussions about the anthropic principle and the multiverse.

2010s
2012:

- **Higgs Boson Discovery:** While not directly related to string theory, the discovery of the Higgs boson at the Large Hadron Collider (LHC) influences the search for supersymmetry and other predictions of string theory.

2015:

- **Gravitational Waves:** The detection of gravitational waves by LIGO opens new avenues for testing predictions of string theory, such as cosmic strings.

2020s
Ongoing Research:

- **Supersymmetry Searches:** Experiments at the LHC continue to search for supersymmetric particles predicted by string theory.
- **Gravitational Wave Observations:** Ongoing observations aim to detect signatures of cosmic strings and other string-theoretic phenomena.

Future Prospects:

- **Next-Generation Colliders:** Proposed future colliders, such as the Future Circular Collider (FCC), aim to explore higher energy scales, potentially revealing new physics predicted by string theory.
- **Advanced Cosmological Observations:** Missions like the James Webb Space Telescope (JWST) and next-generation gravitational wave detectors could provide indirect evidence for string theory.

AFTERWORD

Congratulations! You've made it to the end of our string theory adventure. Take a moment to pat yourself on the back – you've just navigated through some of the most cutting-edge concepts in modern physics.

When we started this journey, string theory might have seemed like an impenetrable fortress of complex math and mind-bending ideas. But step by step, we've we've explored vibrating strings, extra dimensions, and even the possibility of multiple universes. We've seen how this elegant theory attempts to unite the very big (gravity and general relativity) with the very small (quantum mechanics).

Perhaps you're excited by the potential of string theory to revolutionize our understanding of the universe. Maybe you're skeptical about some of its untested predictions. Or you might be a bit overwhelmed by the sheer scope of what we've covered. All of these reactions are perfectly valid!

Even the world's top physicists are still grappling with string theory. It's a work in progress, a theory that continues to evolve as new insights emerge and is still considered a long-term project. The fact that you can now understand the basics of this complex field is a remarkable achievement.

So, what's next? Well, that's up to you! Perhaps this book has sparked a deeper interest in physics, and you'd like to explore further. Maybe you'll keep an eye on news about string theory research. Or perhaps you'll simply enjoy having a new perspective on the nature of our universe the next time you gaze up at the stars.

Whatever path you choose, I hope this book has given you more than just knowledge. I hope it's ignited your curiosity, challenged your perceptions, and reminded you of the wonder and mystery that surrounds us every day.

As we close this book, remember that science is an ongoing journey of discovery. String theory may or may not turn out to be the "theory of everything" we've been searching for. But the quest to understand our universe will continue, driven by curious minds like yours.

Thank you for joining me on this cosmic adventure. Keep questioning, keep learning, and keep marveling at the incredible universe we call home. After all, every scientific revolution starts with a simple question and a curious mind.

Here's to the strings that might tie our universe together, and to the endless possibilities that lie ahead in the vast cosmos.